# CHANGING LANDSCAPES
# OF NORTHWEST INDIANA

Copyright 2025 by Purdue University Press. All rights reserved.

Cataloging-in-Publication Data is available from the Library of Congress.
978-1-62671-146-4 (hardback)
978-1-62671-147-1 (paperback)
978-1-62671-148-8 (epub)
978-1-62671-149-5 (epdf)

Cover: Main Image: Sign describing Beaver Lake installed on US 41 by the Newton County Historical Society. / Michael Dobberstein. Background Image: 1929 drainage map by the US Army Corps of Engineers shows the elaborate network of lateral ditches and dredged creeks used to drain the marsh into the Kankakee River. / US Army Corps of Engineers.

# CHANGING LANDSCAPES OF NORTHWEST INDIANA

DRAINING BEAVER LAKE AND THE KANKAKEE MARSH

MICHAEL DOBBERSTEIN

Purdue University Press, West Lafayette, Indiana

*We of the minority see a law of diminishing returns in progress; our opponents do not.*

*–Aldo Leopold*

# CONTENTS

*Acknowledgments* — ix

Introduction: An Essay on Ownership, and the Question of Value — 1

## PART 1. RESHAPING THE LAND — 11

1. The Kankakee Marsh, the River, and Beaver Lake — 13
2. Selling Beaver Lake, 1853–1891 — 27
3. The "Gigantic Swindle" of 1869–1872 — 51
4. Starting a Current, 1889–1893 — 71
5. The End of the Marsh, 1897–1923 — 87

## PART 2. LAND REMADE: THE CONSEQUENCES — 109

6. Conservationists Raise Their Voices, 1925–1934 — 111
7. Lessons in Hydrology, 1927–1977 — 129
8. Managing the "Monumental Problems" of a Broken River, 1977–1978 — 143
9. Environmentalism Comes for the Kankakee, 1979–1983 — 159
10. Wide Levees and the Struggle for Restoration, 1983–2002 — 177
11. The Kankakee Basin in the 21st Century — 197

*Bibliography* — 211
*Index* — 217
*About the Author* — 231

# ACKNOWLEDGMENTS

A BOOK LIKE THIS CANNOT HAPPEN WITHOUT THE HELP OF MANY PEOple. I first learned of Beaver Lake and the marsh while doing volunteer work at Kankakee Sands in Indiana, where The Nature Conservancy is restoring part of the old lake bed to prairie. Thanks to Alyssa Nyberg for sharing books about the lake from the TNC's library. I would like to thank Beth Bassett, of the Newton County Historical Society, for the Society's newsletters she shared with me, as well as the clippings from the *Kentland Gazette*, and all the advice about Beaver Lake. Thanks to Mary Kay Emmerich, former director of the Newton County Public Library, for guiding me to the histories of Beaver Lake.

I am grateful to John Turner of the Walkerton Area Historical Society for his help in locating nineteenth-century maps of the Kankakee Valley. Special thanks are due to Shirley Hiatt Sherland's invaluable compilation of nineteenth-century Walkerton newspaper clippings in *The Island: Land Between the Kankakee and Pine Creek* (2007), for information about Dixon Place, and his plans to drain the marsh. Andrea Glenn, of the Indiana State Library, directed me to the proceedings of the Indiana Engineering Society, which proved to be key to understanding not only the episode of the rock ledge, but also to the thinking of Indiana engineers about draining the marsh. Dr. Chandramouli Chandramouli, Professor of Engineering at Purdue University Northwest, helped me understand the technicalities of river channelization.

I am indebted to Jody Melton, former executive director of the Kankakee River Basin Commission, and Scott Pelath, executive director of the Kankakee River Basin and Yellow River Basin Development Commission, for information about the commission's work, and for allowing me access to the commission's archives.

Special thanks are due Jim Sweeney for a tour of the river, for his advice about the Izaak Walton League, and for reading part of a draft of this book. Thanks also to David Nichols, editor of *Indiana Magazine of History*, and to my son, Sam, for reading part of the draft. Their comments have made this a better book.

Chapter Two of this book was originally published in the *Indiana Magazine of History* (June 2020), as "The Selling of Beaver Lake, 1853–1889." Chapter Three was originally published in the same journal (September 2023) as "The Gigantic Swindle of 1869–1972." Edited versions are printed here with permission from the editor.

I am particularly grateful to Pat Wisniewski, Jeff Manes, Brian Kallies, and Tom Desch for the documentary, *Everglades of the North: The Story of the Grand Kankakee Marsh*, which is an inspiration for all of us who care about the Kankakee River, and the marsh.

# INTRODUCTION

## An Essay on Ownership, and the Question of Value

*Fellow engineers, keep up the fight; go on conquering and to conquer, until all the subtle energies of nature are overpowered and brought chained captives to do your bidding, and to minister to the necessities and comforts of all mankind.*
—C. G. H. GOSS, PRESIDENT OF THE INDIANA ENGINEERING SOCIETY, IN HIS ANNUAL ADDRESS TO THE SOCIETY, 1895[1]

THIS BOOK IS A STUDY OF THE DESTRUCTION OF THE KANKAKEE MARSH and Beaver Lake in Indiana. The marsh was one of the largest inland wetlands in the country. Beaver Lake was part of the marsh, and the largest lake in Indiana. Both of these remarkable areas were drained in the nineteenth and early twentieth centuries to create farmland. The book offers a detailed history of their draining because the unique characteristics of the lake, the marsh, and the river that was part of the marsh require more than the brief accounts of their destruction that is available in county histories, or in short—and sometimes inaccurate—newspaper retrospectives. For reasons that I hope will become clear, what happened to the marsh, the Kankakee River, and Beaver Lake deserves a proper history. The Great Marsh and the Big Lake did not go easily, or quickly, and for some, at least, these places were incomparably beautiful. For that reason alone, their passing commands attention.

There are compelling environmental reasons for examining the destruction of these wetlands. The first part of the book is a history of the draining of Beaver Lake, the straightening of the Kankakee River, and the consequent draining of the marsh. Draining the marsh not only destroyed a vast and complex ecosystem. This action also caused erosion and damaging sand deposits in the river in Indiana and Illinois, and fostered severe flooding along the Kankakee. Some of these consequences were foreseen by Indiana landowners and engineers in the nineteenth century, before the marsh was fully drained. In the twentieth century, these problems caused bitter and prolonged disputes between residents of the two states. The second part of this book examines the environmental and political consequences of Indiana's transformation of the Kankakee Marsh.

The river that defined the marsh, and was transformed to drain it, flows from Indiana into Illinois, where it joins with the Des Plaines to form the Illinois River, a tributary of the Mississippi. The Kankakee begins in Indiana, but the character of the river changes abruptly in Illinois. An Illinois Department of Natural Resources report summarizes the difference:

> In Indiana the river system has been constructed and managed as an agricultural drainage project—successfully draining the wetlands and converting them into very productive agricultural land. The intent of the management has been based on the economics of agricultural production. In Illinois, especially in Kankakee County [which borders Indiana], the river has been used as a scenic, cultural, and recreational resource.[2]

To drain the marsh, the Kankakee River was straightened and deepened, turning a once sinuous watercourse into a man-made drainage ditch. In Illinois the river retains its ancient winding, meandering journey on its way to join the Des Plaines. These two rivers, the straightened channel in Indiana and the natural stream in Illinois, reflect different attitudes toward the Kankakee, but they also reflect different ideas about how humans relate to the natural world. In Illinois, the Kankakee has been dammed in places, but it has been left largely alone and has retained its original character. In Indiana, a huge wetland was destroyed, the largest lake in the state obliterated, and a river converted into a ditch.

True, the Kankakee was easier to manage in Illinois. Once it crossed the state line, it behaved like a river should. It had a current, it was navigable, and it had banks. Towns like Momence and Kankakee could be built beside it. In Indiana, the river flowed through a broad, flat, sandy plain, where for eighty miles there was very little difference in elevation. The river barely had a current, and in many places it had very low banks. Vast marshes extended for miles north and south of the river, and many in Indiana saw the river and the marsh as one entity. In 1861, an article in an Indianapolis newspaper described the problem succinctly, with only a little exaggeration: "It is all one swamp, and all one river. The river is only a little deeper place in the swamp, or the swamp is only an extension of the river. There are no banks, no limit to mark the separation of one from another."[3]

Indiana officials knew what they were up against. What could the state do with a huge expanse of territory that had no obvious resources? Most of the river in Indiana was not navigable. Towns could not be built in a marsh. More importantly, farming was impossible and land that could not be cleared for crops had no value. It was useless. To many, the problem was that nature was malevolent; they saw the marsh and the river as "*injuring*" the land.[4] Such language implies not only that land and the marsh were separate, but that the marsh, somehow, was the enemy of the land, and wished it harm. The "harm" was that marshland could be sold for only three dollars an acre, whereas productive farmland could be sold for twenty dollars an acre.[5] To extend the logic of this metaphorical injury to the land was to conclude that the marsh and its river were robbing the state of millions of dollars.

Beaver Lake was about five miles south of the river near the Illinois state line. Surrounded by wetlands, it was geographically part of the Kankakee Marsh. The lake became the property of a speculator who acquired title to it in a questionable manner as a temporary investment, a means to eventual profit. Ownership confers certain rights, after all. We believe that we can do whatever we want with our property, and the most obvious right of ownership is the right to sell that which we own. Speculators, of course, want their property to increase in value, and one way to ensure that is to "improve" the land. Put up a fence. Build a house, plant a garden, dig a well. If your property happens to be a lake: drain the lake.

Nothing that happened in the past was inevitable. The citizens of Indiana, its legislators, investors, engineers, and landowners made choices about how to manage the Kankakee River lands in their possession, choices based on beliefs not only about the use of land, but on beliefs about the nature of ownership, and about how to determine the value of property, especially land. In the last century or so, a shift in social outlook has shaped government policies that focus on regulation to protect the environment from the worst depredations caused by the belief that land has only utilitarian or economic value. A premise of this book is that social policies, though essential to preserving natural areas, are not enough. Value beliefs of individual property owners are crucial in determining the fate of natural environments. This idea is expanded in Chapter 11.

We think about value in different ways. One kind of value is economic. One's property has a market value, a price that can be allotted to it. This kind of value is adjustable, depending, for instance, on how much demand might exist for property at a given time. Economic value is calculated in terms of money. Sometimes, value has a much broader application. We may speak of a person's "values," and these attitudes, or beliefs, or "ways of looking at the world" are culturally based and may be completely irrelevant to matters of cost or price. These values determine how we think about the world, including nature, and they determine how we relate to others, and to the natural world, the world of trees and wetlands and oceans and mountains. These kinds of values determine behavior. How we think about nature, of course, is bound up with our notions of economic value, because property is often land—the natural areas of the world. This book advocates for conservationist values. That is, I propose that we consider that nature may have a value that transcends the economic; indeed, nature may have a value that transcends the cultural, the merely human. Nature may have a value all its own. This idea is explored more fully in the last chapter.

This is not a prominent idea in western thought. It is, in fact, a recent idea, only very recently widely advocated. This book is not a work of philosophy, it is a work of history, and I believe it will show how certain values are imbedded in history, which means imbedded in culture, and implanted in all of us today. The people who drained the marsh, or paid for draining the marsh, or created laws that enabled its draining, or chartered organizations for draining, or who merely stood by and watched with approval would all have agreed with the sentiments expressed by engineer C. G. H. Goss at the beginning of this introduction: that nature should be made a slave to humans. Such an idea was unquestioned by most—though by no means all—at the time the marsh was drained, and since I am proposing an alternative to those values, it is worth discussing their philosophical context.

The idea that nature should be a slave to humans is linked to our beliefs about ownership, and our ideas about property. No one expressed the principles that underlie the western values of property better than John Locke, a seventeenth-century English philosopher who thought a lot about ownership, about what we might call the value of property, and about the role of nature in those values. Locke's ideas were revolutionary for his time because he taught that people's rights were a condition of being human; that is, each individual was born with them. Therefore, those rights took precedence over the powers of government, which in Locke's time was headed by a king.

Locke's ideas would be familiar to most Americans today, and certainly to the citizens of earlier times. Thomas Jefferson grounded some of his ideas in Locke's writings. Every American is familiar with the famous phrase in the Declaration of Independence that grants all of us a right to "life, liberty, and the pursuit of happiness." Locke uses the phrase "pursuit of happiness" in his book, *An Essay Concerning Human Understanding*, and connects the pursuit of happiness with the foundation of liberty.[6] Locke says in his *Two Treatises of Government* that people have a fundamental right to life, liberty, and estate (property), meaning that these rights do not depend on government. In the second *Treatise*, he devotes an entire chapter to the subject of property.[7] Locke believed that we own things as a consequence of being human. We each have a property in our own person, he wrote. The labor of our bodies belongs only to us, thus we own what we have worked for—have a right to it, in fact—because labor, combined with the products of nature, constitutes our property.[8] Locke was proposing a theory of *private* property based on nature—and that our right to that property is as fundamental as our right to life.

Locke was thinking about America. He uses native people as examples of his idea of property as labor, writing for instance that "the venison, which nourishes the wild Indian" must belong to the Indian, and to him alone. Locke says that the "law of reason" allows the Indian to own the deer that he has killed, and he seems to conceive of aboriginal America as a place where one might imagine the primordial conditions surrounding the origins of property rights.[9]

Locke lived during a time when Europeans were beginning to colonize America, and he does not discuss the effects of colonization on native people. But we might ask: Who owned first what we own now? The original inhabitants of northern Indiana—the location of the Kankakee Marsh—were the Potawatomi, a tribe that also lived in southern Michigan, Illinois, and Wisconsin. In 1832, the Potawatomis of northern Indiana were forced to cede most of their territory to Indiana. That cession, and the idea of the treaty, implies that the Potawatomis owned the land, and had the right to sell it. However, the Potawatomi, like most tribes, conceived of "ownership" in quite different terms than settlers, and did not share the concept of privately owned land that could be traded for money. Most natives saw land as a commons, whose use was negotiated among many people. To speak of tribal "ownership" of land, therefore, always requires qualification.[10]

Most of the Potawatomis moved to Kansas, but a few stayed in northern Indiana on reservations determined by treaty. In 1838, the governor decided that the state needed the small pieces of land left to the tribe. The last of the Potawatomis, led by Chief Menominee, had to go. Menominee claimed that his signature on the agreement giving up his reservation had been forged, and that other chiefs had been drunk when they signed. The governor sent a force of one hundred armed men to remove Menominee and his band of holdouts. Over eight hundred people were forced to march to Kansas. Threatened with bayonets and crowded into wagons, over thirty died of heat and suffocation—most of them children.[11]

Locke believed that people would not dispute ownership of land when they had enough for their own needs. "What portion a man carved to himself was easily seen; and it was useless, as well as dishonest, to carve himself too much, or take more than he needed," he wrote.[12] This idea seems naïve, but Locke understood that when money was involved, people would want to "enlarge their possession."[13] In other words, it's just human nature to want to get rich. The desire to enlarge one's possession—to take more than is needed—is an important motivator in American history.

History has many lessons about ownership, and about property. I live in a town in northwest Indiana that was founded in 1866 by a German immigrant named Nicholas Scherer who found work draining local marshlands, and eventually in real estate. By 1874 he had enough land of his own to give some away, so he donated a few acres to his religion, for a church.[14] His is among the oldest graves in the cemetery next to the church of St. Michael the Archangel. (The church enforces property rights, even for the dead. A sign at the entrance to the cemetery says, "Church Members Only. No Trespassing.")

The German immigrant bought the land for the town from a farmer named Aaron Hart who owned thousands of acres that he had bought from the government, cheaply. The farmer was a land speculator. Like Scherer, who bought property from him, then sold it, Hart bought land believing that its value would increase, and he could make a good profit when he sold it. Locke said that unimproved land is wasted, that the value of property can only be calculated by the uses to which we put it. Land that remains wilderness has no value, he said.¹⁵

Aaron Hart worked hard at improving his property, no doubt with a sharp eye toward improving its sale price. I don't know if Hart ever read Locke, but I am sure he would have agreed with the unequivocal statement in Section 43 of the *Second Treatise of Government* that says that "it is labour then which puts the greatest part of value upon land, without which it would scarcely be worth any thing." With that observation, Locke becomes almost lyrical on the subject of labor. It is labor, and labor alone, he says, that transforms nature into something of consequence, something of value:

> [F]or it is not barely the plough-man's pains, the reaper's and thresher's toil, and the baker's sweat, [that is] counted into the *bread* we eat; the labour of those who broke the oxen, who digged and wrought the iron and stones, who felled and framed the timber employed about the plough, mill, oven, or any other utensils . . . must all be *charged* on the account of labour, and received as an effect of that: nature and the earth furnished only the almost worthless materials.¹⁶

The land that Hart bought from the government was mostly swamp—uninhabitable and useless for crops, much of it underwater. He designed a network of ditches to drain the swampy land into nearby creeks and rivers. It was hard work. At one point Hart employed forty men who drove fourteen yoke of oxen, two teams of horses, and three teams of mules. It took him nearly twenty years, but by the 1880s he had transformed a swamp into farmland, and into land for towns. At the time of his death he owned seventeen thousand acres and had made himself rich. Before moving to Indiana, he had been a publisher in Philadelphia, but he seemed to relish the hard manual labor of digging the swampy earth. He was working at the bottom of one of his ditches on a January morning in 1883 when its walls collapsed. He died instantly.¹⁷

Hart's death reminds us of the heavy price that labor can exact. The death of Potawatomi children ought to remind us of the incalculable cost of the desire for territory, for more property, to take "more than is needed," in Locke's terms. How should we "charge" these costs? To what account can we reconcile this loss? After the last of the Potawatomis were forced out, ownership of land was consolidated under the government of Indiana, which sold some to Aaron Hart, who sold it to the German immigrant, who sold it to developers, who built a town. Eventually, a piece of it was sold it to me.

This bit of history offers a plausible answer to the question, "Who owned first what we own now?" Ownership of land as we understand it—as Locke would have understood it—only has meaning within a particular frame of reference in which a territory is mapped, surveyed, measured, parceled out, and sold. Measurements must be accurate and real estate sales require title certifications. Buyers must be assured that owners have a legal right to sell; claims to the land must be valid. These are the ethics of ownership, enshrined in custom and statute. It is doubtful that the Potawatomi—or any other tribe that held land in common—believed in, or understood, this system of private ownership. The native view of land as a common living space was never considered as an *alternative* to the idea of private property. Colonists simply saw native lands as unowned because they were not "improved"—raw nature had not been transformed into anything of real value, in Locke's terms.

The system of private property measures land, carves it into sections, and sells it piecemeal for money. That observation invites the question: what is money, and how has it become the measure of value? Locke had some thoughts about money. He believed that the value of gold and silver—money—derived not from labor combined with nature, but only from the "consent of men."[18] In other words money, otherwise worthless, becomes a measure of value only because of an agreement among humans. Locke thought that money allowed people to enlarge their possessions beyond what they needed. This consent, he wrote, leads to another: that men "have agreed to a disproportionate and unequal possession of the earth."[19] This agreement, it seems to me, is a tacit approval of greed.

I have quoted Locke at length because I think he is a convenient reference point in this discussion. His writing crystallizes ideas about ownership that have been current for some time. Locke's belief that nature has no intrinsic value, that it is worthless until shaped into something useful by humans, is shared by many Americans. Unlike most, however, Locke was dubious that money conferred real value on anything. However, by denying nature any value of its own, he failed to appreciate how easily money would become the common measure of value set upon land as well as labor, and how greatly that would strengthen the desire for a disproportionate and unequal possession of the earth.

Locke's ideas about private property are the foundation of the modern idea that ownership of land confers an unquestioned right to use that land solely for private gain. However, this is not the only tradition of the function of property in American history. Lynn Heasley, in her study of shifting views of ownership in rural Wisconsin, reminds us of another view, inspired by Thomas Jefferson, that maintains "that property is a tool for ensuring the public good and, ultimately, for securing democracy." Heasley argues that "democratic societies have a fundamental interest in moderating the influence landowners have over public policy; otherwise their disproportionate, or concentrated power puts democracy at risk."[20] Not only democracy is at risk, however. The theme of this book is that unrestrained influence of landowners over public policy not only can damage natural environments, it endangers the property of other landowners. Indeed, it can put entire communities at risk.[21]

This study provides an object lesson in the danger of narrowly privileging individual rights over the common good. The government of Indiana consistently enabled landowners to treat public waterways such as Beaver Lake and the Kankakee River as though they were private property. This policy caused the complete annihilation of the largest lake in Indiana, the almost complete destruction of one of the country's largest inland marshes, and the environmental degradation of one of the state's major rivers. The loss of a major wetland area and the channelization of the river have enabled a condition of permanent annual flooding throughout the Kankakee Valley.

Legislators and landowners in early twentieth-century Indiana set out to transform nearly half a million acres into farmland, not primarily to rid the land of pestilential swamps, or to feed growing numbers of hungry citizens in the state, but in order to make money. By the 1850s, railroads had come to northern Indiana, and legislators and landowners alike saw the potential of the Chicago market for farm produce. Land became, not as in previous times, a space for subsistence farming and communities, but rather a form of capital, something to be used, and used up, for the sole purpose of increasing wealth.[22] The potential value of each acre in the Great Marsh was arrived at by a calculation as simple as the one made by the *Indianapolis Sentinel* in 1861: farmland was worth ten times more dollars than marshland. The simple logic of the market required that the marsh be destroyed. This book is a chronicle, not only of the events of the marsh's destruction, but of what happens to the natural world when we absolutely believe in the values articulated by engineer Goss and John Locke.

The economic forces that descended upon the wetlands of northern Indiana in the last half of the nineteenth century were part of the economic revolution that changed the landscape of America. Historian Carolyn Merchant summarizes the profound changes that the core beliefs driving this revolution brought to society, and to the environment:

> It threw open land, water, air, and all the life they contained to unrestrained development in the pursuit of wealth and status. It made profit-and-loss the sole criterion for dealing with nature, conceived as inert matter. A mechanistic worldview based on the quantification of matter and energy, interchangeable parts, mathematical prediction, and the control of nature replaced the animate cosmos of the colonial farmer. In its wake lay cut-over forests, smoky air, polluted streams, and endangered wildlife.[23]

John Locke—or for that matter Thomas Jefferson—could not have imagined the power unleashed by the market economy that Merchant refers to. But Locke had formulated the foundational principle of this destructive force: that nature in itself is worthless—"inert matter," in Merchant's useful phrase. Nature conceived of in this way is only a commodity to be processed, or an impediment to be removed. It cannot be truly valued, only evaluated in relation to the market. Something that is worthless and inert is not beautiful, nor can it be loved. To put this another way: that which is loved, which is considered beautiful, cannot be worthless. It is, in fact, priceless. There were those who loved the river and its marshes, and Beaver Lake. They found these places surpassingly lovely. This book is dedicated to their memory.

# NOTES

1. C. G. H. Goss, "Address by President Goss," in *Proceedings of the Fifteenth Annual Meeting of the Indiana Engineering Society, Held at Indianapolis, IN, January 7, 9, and 10, 1895* (Indianapolis, 1895).
2. J. Loreena Ivens et al., *The Kankakee River Yesterday and Today* (Illinois Department of Energy and Natural Resources, 1981), 2.
3. "A Chance for Speculation," *Indianapolis Daily Journal*, April 2, 1861.
4. Emphasis the author's. See "The Kankakee Valley Draining Company," *Indiana Weekly Sentinel*, April 3, 1861; and *The Kankakee Valley Draining Company Prospectus, 1869* (Newton County Clerk's Office, Kentland, IN), 7, for examples of this strange usage, which was often employed to show the cause for rendering the land "valueless." All newspaper articles, except as noted, were accessed online at Newspapers.com or Newspaperarchives.com.
5. "A Chance for Speculation," *Indianapolis Daily Journal*, April 2, 1861.
6. *John Locke, An Essay Concerning Human Understanding*, vol. 1, collated by Alexander Fraser (Oxford, 1894), Book II, Chapter XXI, Section 52.
7. *Locke, Two Treatises of Government*, prepared by Rod Hay for the McMaster University Archive of the History of Economic Thought (London, 1823), Book II, Chapter II, Section 87.
8. Locke, *Two Treatises*, Book II, Chapter V, Section 26.
9. Locke, *Two Treatises*, Book II, Chapter V, Section 29.
10. Thomas Campion, "Indian Removal and the Transformation of Northern Indiana," *Indiana Magazine of History* 107, no. 1 (March 2011): 40. See also William Cronon, *Changes in the Land: Indians, Colonists, and the Ecology of New England* (Hill and Wang, 1983), 54–81, for an extended discussion of the differences between native and colonist views of land and property.
11. Irving McKee, "The Centennial of the Trail of Death," *Indiana Magazine of History* 35, no. 1 (March 1939): 38–39.

12. Locke, *Two Treatises*, Book II, Chapter V, Section 51.
13. Locke, *Two Treatises*, Book II, Chapter V, Section 49.
14. Art Schweitzer et al., *Schererville Through the Years: A Pictorial Look Back* (Schererville Historical Society, 2002), 3.
15. Locke, *Two Treatises*, Essay II, Chapter V, Section 42.
16. Locke, *Two Treatises*, Section 43.
17. T. H. Ball, "Aaron Norton Hart," Lake County IN Archives Biographies, http://usgwarchives.net/in/lake/bios/bios-h.html.
18. Locke, *Two Treatises*, Book II, Chapter V, Section 46.
19. Locke, *Two Treatises*, Book II, Chapter V, Section 50.
20. Lynn Heasley, *A Thousand Pieces of Paradise: Landscape and Property in the Kickapoo Valley* (University of Wisconsin Press, 2005), Kindle Edition, location 198.
21. Illinois has long claimed that Indiana's agrarian policies on the Kankakee have damaged the river in Illinois. A particularly egregious example of threats to communities is the creation of large algae blooms in waterways that furnish water to populated areas. Note the algae bloom in Lake Erie, caused by fertilizer runoff from farms, that contaminated the water suppy of Toledo, Ohio, in 2014. See "Toxic blooms on Lake Erie still a problem 10 years after Toledo issued a 'do not drink' order," *Michigan Public*, August 2, 2024, https://www.michiganpublic.org/environment-climate-change/2024-08-02/toxic-blooms-on-lake-erie-still-a-problem-10-years-after-toledo-shut-down-its-water-system. Also annual gigantic algae blooms in the Gulf of Mexico, caused by agricultural runoff from farms into the Mississippi River, have created a "dead zone," threatening the fishing industry in that area. See "What Causes the Dead Zone?," The Nature Conservancy, last updated October 25, 2024, https://www.nature.org/en-us/about-us/where-we-work/priority-landscapes/gulf-of-mexico/stories-in-the-gulf-of-mexico/gulf-of-mexico-dead-zone/.
22. William Cronon makes this point about colonists in the seventeenth century, as they transformed the land of New England into a capitalist economy in *Changes in the Land: Indians, Colonists, and the Ecology of New England*, 169. This transformation swept the American landscape in the eighteenth, nineteenth, and twentieth centuries, and still does.
23. Carolyn Merchant, *The Columbia Guide to American Environmental History* (Columbia University Press, 2002), 68.

# PART 1
# RESHAPING THE LAND

I

# THE KANKAKEE MARSH, THE RIVER, AND BEAVER LAKE

Hunters, Historians, a Geographer, and a Poet

B Y THE TIME THE LAST OF THE POTAWATOMIS HAD BEEN REMOVED FROM the state, the tribe had lost three million acres in northern Indiana. This area included Beaver Lake, the Kankakee Marsh, and all of the Indiana portion of the Kankakee River. The Kankakee drains around a million acres in Indiana alone, and the marsh was huge, comprising the river, Beaver Lake, and four hundred thousand acres on either side of the river. Before it was straightened, the winding river in Indiana was over two hundred miles long, flowing a linear distance of just over eighty miles, from just south of South Bend to the Illinois state line. The marsh was six to ten miles wide on either side of the river, depending on the season, and rainfall.

Beaver Lake was located in what is now Newton County, about five miles south of the Kankakee, near the Illinois border. The lake was about eight miles wide (east to west) and five miles long (north to south). Almost completely surrounded by Kankakee wetlands, it was a geographical part of the marsh. It was a big lake, covering about sixteen thousand acres. The total lake area, including its surrounding marshes, was thirty-six thousand to forty-six thousand acres.

There are books about the marsh, written by people who witnessed its destruction. More accurately, there are anecdotal histories about people who lived in or near the marsh. Fay Folsom Nichols's *The Kankakee: Chronicle of an Indiana River and its Fabled Marshes* recounts history and legend, sometimes fictionalizing both. Despite the reference to the marsh in the subtitle, the book contains little information about the actual marsh. Nichols provides a brief summary of the draining projects, which she condemns strongly, but gives few details.[1] J. Lorenzo Werich's *Pioneer Hunters of the Kankakee* is both memoir and history, but Werich was a hunter and a trapper, and he focuses on those who shared his interests. Interestingly, he challenges the idea that the marsh was "useless." He points out that between 1850 and 1900 over three million dollars in furs were taken from its waters.[2] Earl H. Reed, in *Tales of a Vanishing River* (1920) mourns the passing of the marsh, commenting accurately that "the utilitarian has triumphed over beauty and nature's providence for her wild creatures. The destruction of one of the most valuable bird refuges on the continent has almost been completed, for the sake of immediate wealth."[3] Reed points out that his book is not a history, but rather "an interpretation of life along the river."[4] He fictionalizes stories of early settlers and hunters in the marsh, but like Nichols and Werich his emphasis is on people, not the marsh itself.

In the following pages I quote liberally from several sources that describe the marsh and the lake. These sources are scattered and fragmentary. Only a small fraction of the marsh remains, and Beaver Lake has been completely erased. Neither had an Ansel Adams to capture them in lustrous photographs, or an Albert Bierstadt to render them as vibrant, sweeping landscapes. The marsh and Beaver Lake were major stopovers for thousands of migrating waterfowl, but no study exists of their numbers, or their species. No scientific accounts of other wildlife or the once exuberant plant life of either the lake or the marsh were ever written while these places actually existed.[5] "Official" histories say little about how these unique areas actually *looked*, or about what life forms they contained. The few firsthand accounts we have that provide specifics about their incredible diversity of wildlife and plants are moving testimony to what was lost.

A good overview of the Kankakee River, and of the marsh, is from Joel Greenberg's *A Natural History of the Chicago Region* (2002):

> For the first two-thirds of its length, the 110 miles from South Bend, Indiana, to Momence, [in Illinois, about 6 miles from the state line], the [Kankakee] river originally twisted into two thousand bends as it dropped in elevation just a little more than a hundred feet. (With the 250 miles of meanders, this decline averaged about five inches per mile.) The sluggish water rolls over a thick bed of gravel and sand, until it reaches a four-mile-long limestone outcropping or ledge at Momence. This divides the river basin in two and acts as a natural dam to impound the drainage of the upper valley, a watershed exceeding two thousand square miles. The natural reservoir cradled the "Great Marsh," one of the largest fresh-water wetlands in the United States ... For most of the year, sheets of water three to five miles wide and one to four feet deep flanked each side of the river channel.[6]

In 1936, in his dissertation for a PhD in geography, Alfred H. Meyer provides a topographical summary:

> Marsh prairies of aquatic sedges and grasses, potential grazing areas; wild-rice sloughs, scenes of countless wild geese and ducks; flag ponds, lined with muskrat houses; a narrow but almost uninterrupted swamp forest, full of game, rimming a meandering river teeming with fish; the wet prairies, made humanly

habitable by the interspersion of sandy island oak barrens surmounting the highest flood waters.[7]

Meyer was from Illinois, but born in 1893, he probably never knew the marsh firsthand. As Greenberg points out, Meyer is alone in devoting scholarly attention to the marsh.[8] His dissertation is a study in land use, a popular topic in New Deal policies of the 1930s. In the section on sportsmen's use of the marsh, he betrays his conservationist leanings when he interrupts the dry listing of hunting clubs with this imaginative evocation:

> Reconstructing the former scene of the sportsman's Kankakee, we observe that the swamp rimming the river is sufficiently narrow to act only as a momentary screen to the completely changed yet hardly less charming view of the marsh without. Here, as far as the eye can see, is an expanse of marsh interrupted only by an occasional upland oak grove on a sandy knoll or ridge, or possibly by a cluster of closely set pin oaks rising out of the water, with their characteristic scraggy downward-bending dead branches. Miniature mounds with clumps of grass just emerging from the water betray the somewhat undulating character of the otherwise markedly far-flung flatness of the submerged terrain. Rank sedges and grasses (source of marsh hay) and flags mark the shallower stretches. With these alternate the wild-rice swales; the cattail, spadderdock, and smartweed slough; and the reed and lily ponds, water sites throughout the year. Domes of earth and flags, the homes of whole colonies of "rats" [muskrats], dot the rims of sloughs, ponds, and poorly defined stream channels. The wild-rice and smartweed fields are the scenes of countless geese and ducks and other wild fowl which find here their ideal harvest by day, and in the neighboring swamp their haven by night.[9]

Meyer should not be faulted for inserting a bit of creative writing into his dissertation. By the mid-1930s when he was finishing his PhD, the Kankakee Marsh was almost completely gone. Physical descriptions of it were hard to find, and those who had first-hand experience of living and hunting there were fast disappearing. Meyer reinforces his imagined scene with eye-witness reports from two Indiana hunters:

> In the spring, ducks generally began to arrive around the middle of February, or as early as February 1. Mallards and pintails were first to arrive, followed by teals and bluebills, then spoonbills, and finally the wood ducks, which came about the middle of March. Wood ducks often fed on the timber acorns; others were strictly marsh birds and would not go into the timber swamp except during a storm. Sand Hill cranes and brant [geese] were plentiful in early spring, and later came the jacksnipe,

plover, and rail to stay till the water seeped through the sand on the higher marshes, where the prairie chicken could be heard calling at break of day.[10]

Hunters were good observers, and sometimes, very good naturalists. F. E. Ling, an Indiana dentist who hunted and fished in the marsh, remembered scenes of great natural beauty:

> Very many kinds of trees, shrubs, flowers, grasses and aquatic plants grew in this area. Many of the bayous and pond holes in the timber and marsh were a white carpet of water lilies and the banks of the bayous and ponds were lined with marsh hollyhocks [hibiscus]... In the spring the islands were one mass of flowers—spring beauties, Dutchman beeches and violets. Later the May apple covered large areas with its umbrella foliage and the Indian turnip with its green calla lily bloom was seen everywhere.[11]

Ling had a keen eye for trees and shrubs as well as flowers:

> The trees on the islands were beech, white oak, burr oak, red oak, black oak, hickory, pepperage, butternut, black walnut, white ash, black ash, sycamore, soft maple, sassafras, wild cherry and paw-paw. On the lowlands there were pin oak, burr oak, black oak, red birch, elm, black ash, white ash, sycamore, cottonwood, soft maple, quaking aspen and willow. Acres of witch hazel, blackberries, raspberries, blueberries, huckleberries and wild strawberries were found on the islands and around the edge of the marsh. Large areas of puckerbrush and devil pins were found in the lower areas around the bayous and ponds.[12]

Some of the best writing about the marsh is from Charles H. Bartlett, a high-school principal from South Bend. Born in 1853, Bartlett grew up near the eastern edge of the marsh, and in 1907 published *Tales of Kankakee Land*, which combines personal reminiscence and re-telling of stories that he heard as a boy. He begins his book with a romantic overview of the marsh:

> More than a million acres of swaying reeds, fluttering flags, clumps of wild rice, thick-crowding lily pads, soft beds of cool green mosses, shimmering ponds and black mire and trembling bogs—such is Kankakee Land. These wonderful fens, or marshes, together with their wide-reaching lateral extensions, spread themselves over an area far greater than that of the Dismal Swamp of Virginia and North Carolina. Their vastness, their silence, their misty haze, and their miry depths make them the very realm of forgetfulness and oblivion.[13]

Bartlett exaggerates the size of the marsh, which was probably smaller than the Great Dismal Swamp (a great deal of which still exists as a national wildlife refuge and a state park—unlike the Kankakee Marsh), but its huge size invites the comparison. In a later chapter, he provides a poetic description of an "island" called Eagle Point, one of the numerous sandy hillocks that rose above the marshy ground:

> This island, whose quiet haunts we loved to invade, was covered in most parts with an oak-grove, with here and there a giant shell-bark hickory. The soft turf spread beneath the grove was screened from view on all sides by the tops of dense thickets of dogwood, and marsh maples and soft willows that rose from the low ground surrounding the island, their upper branches glancing over into the higher plain which they could not invade. Here and there, over the interior, was a clump of sassafras or a billowy area of wild roses. There was a place where a few white birches lifted their graceful, though ghostly forms—a rear-guard of the forest flora that flourished here at the close of the ice age... Where a boggy indentation at the base of the island completed the latter's heart-shaped outline, there stood a dark, compact mass of tamarack. The delicate foliage of their tender green rose in exquisite contrast against the dull gray wall of massive oak-trunks that leaned from the top of the bank and far on high spread out their leafy branches, as with solemn invocation of peace.[14]

Neither hunter, trapper, or naturalist, Bartlett emphasizes the aesthetic possibilities of close observation. This lyrical passage, in fact, is almost Wordsworthian in its attention to visual detail that resonates with a heightened awareness. This is consistent with the tone of the entire book. Bartlett not only wants to describe a scene, he wants to convey an intense, subjective experience. By the time his book was published in 1907, at least half the marsh was gone, including the area he describes in this passage. The book does not mention this—nor does Bartlett acknowledge the imminent destruction of the rest of the marsh, which he surely must have realized was inevitable. Perhaps he did not want to threaten his memories by acknowledging the disappearance of what was remembered, and loved.

The Kankakee Marsh spread into eight different Indiana counties, yet most nineteenth-century and early twentieth-century county histories provide no real description of it. Some barely mention the marsh. A 1907 history of St. Joseph County (where Charles Bartlett grew up), for example, notes that "drainage and fencing has robbed [the marsh] of its original charm," but does not explain further.[15] The history of Starke County devotes an entire chapter to an account of the drainage of the marsh, extolling the beneficial effects of the change in the terrain, but contains no description of the marsh at all. The writer proudly emphasizes the extensive effort required to transform the geography of the land, noting that "all the natural streams in the county have been dredged, widened, straightened and deepened . . . the work going on almost every day and night the year round."[16]

Most histories want to find something "useful" in the marsh, such as the 1880 history of LaPorte County, which quotes a geologist's report that iron ore had been discovered there. The report speculates that the peat of the marsh could be converted to fuel for a blast furnace, so that marsh and ore "may add to the value of the other, and naturally tend to bring the much-abused Kankakee marsh into more favorable notice."[17] The 1904 history of LaPorte County actually denies that the marsh even existed, saying that the "first settlers in the county, and until recently the later settlers have been deceived in this matter." The writer claims that the Kankakee Valley, which contains many acres of marsh grass, has "no fen, bog, morass, swamp or quagmire."[18] A 1912 history of Porter County mentions only the commonplace fact of the time, that the marsh had some of the richest soil in the state, "a dark, sandy loam, rich in organic matter," whose cultivation brings the "universal result" of good crops.[19] The 1916 *Standard History of Newton and Jasper Counties* devotes over four pages to the litigations surrounding the draining of the marsh but does not describe it.[20]

A notable exception to the general disinterest in any physical description of the marsh is the 1915 history of Lake County, which devotes a page to a brief inventory of its plant life and suggests that the marsh had a value in itself:

> For years it was the paradise of the white and the yellow lily and the cattail, as well as the blackbird, the bobolink and the muskrat. The cranberry was also a native of the marshes. The swamps also had quite a timber growth of ash, elm, sycamore, birch, willow, maple and cottonwood, while on the islands, which are generally sandy, were clusters of oak, hickory, sycamore, beech, walnut and maple. Most of the wooded tracts in the Kankakee marsh are in the southeastern corner of the county, as many as six sections in that region being originally covered with timber, mostly with ash and elm, with some sycamore and gum trees.[21]

Beaver Lake, the largest lake in the state when Newton County (then Jasper County) was settled by whites, also attracted little interest from local historians. John Ade's 1912 history contains a brief chapter about the lake, but after providing its location and size, Ade says only that "myriads of geese, ducks, swan and other game birds would be found there."[22] He provides a brief summary of the draining of the lake and concludes, "[t]he name remains, but the 'lake' itself is now only a memory of the past." The 1916 *Standard History of Newton and Jasper Counties* contains a valuable account of the disputed title to Beaver Lake (more of that in the next chapter) but says little about what the lake actually looked like, beyond a cursory note about the variety of game and fish that lived there.[23]

It remained for historians of a later era, after the lake was destroyed, to collect stories about actual life on its waters. In his 1925 book about the Kankakee River and Beaver Lake, Burt Burroughs remarks: "How is one to go about it to tell the interesting story of this moving picture of wild life of the long ago?" Burroughs quotes Judge Hunter, who hunted and fished on the lake, in answer:

> Man! Man! Man! The spectacle was too stupendous for words! He would had to have known something of those days; he should have lived in the swamps as I did, weeks and months at a time; he would have to have the echo of the deafening clamor of all this wild life in his ears, and envision the gray-white bodies and yellow legs of these mighty hosts all set to drop into the open water spaces among the rushes and wild rice!

Burroughs wrote about waterfowl that occupied the lake by the acre:

> At this time small, red-head teal occupied these waters literally by the hundred of thousands. Out on the broad expanse of the lake proper and the open reaches of that famous hunting ground known as the "Gaff Ranch," there dwelt mallards, geese, and swans literally by the acre. Judge Hunter recalls that often as he lay in his blind he has watched a flight of these red-heads go over, scarcely six feet above his head, a veritable cloud of them acres in extent, a living blanket four or five feet in thickness.[24]

Burroughs says that Joseph Kite, a resident of Indiana near the lake, maintained that

> cold, hard figures, even though one employs the term thousands, or hundreds of thousands fail to adequately express the idea of unlimited numbers of wild fowl that occupied the waters of the lake. Beaver Lake contained roughly speaking, thirty-five to forty thousand acres, mostly covered with water [counting the marsh area around the lake]. Therefore, he would use the term "acres," as most to express of numbers of the mallards, geese, brant, and swan that frequented the place. The swan especially were numerous.[25]

Elmore Barce, an Indiana lawyer and historian, was also interested in collecting stories from those who had lived near the lake. In 1938, he published *Beaver Lake, Land of Enchantment*, which contained reminiscences of Ned Barker, whose father had built a house on the lake in 1842. Barker recalled seeing "a hundred acres of swans at a time. At a distance they looked like a white island and would remain in the same locality for a month at a time, feeding on the water celery."[26] Barker remembered thousands of Canada geese taking flight from the lake:

With the first streaks of the dawn—at a time when the distant lines of the timberland are scarcely discernible—the call of the assembly is sounded, the honk of the gander, answered by the eager and expectant cries of the feathered hosts. From widely separated positions on the face of the lake, the sudden ascent and 'take off' of the leaders begin. There is a long flap over the surface of the wave, a loud splashing with feet and wings, and the heavy body of the gander, like a rising plane, mounts the first barriers of the air and sweeps in a long, graceful curve to the heights beyond. Hundreds—thousands—follow in his wake, and soon there is the rising of bodies and beating wings, a constant vibration of the air, that rattles the window-panes two and three miles inland from the shore.[27]

The most poignant, and to my mind the most moving description of Beaver Lake comes from Indiana judge Charles Test, who camped at Beaver Lake with his sons in the 1850s:

Long before sunrise I left the camp to bathe my face and hands in the waters of the lake. The scene was changed with the dawn of another day. The rising mist just touched the tranquil lake, and the refreshing morning air gave new vigor to the human frame. Not a leaf stirred. Not a sound came from the forest or lake. It seemed indeed that the very spirit of tranquility had spread her downy wings over all, and hushed the unearthly sounds of the previous night. The fish again leaped from the surface of the water. What a change. So quick and so bright. Surely the magician's hand was invisibly shifting the scene. The great orb of day lifted his disk above the hills, and poured his glory into the darkness across the lake and into the forest beyond. The sublime expression of the sacred historian describing the beginning of the world rushed across my memory: "And the evening and the morning were the first day."[28]

Before Indian removal, the marsh and Beaver Lake were home to Potawatomis as well as white settlers, traders, and hunters. Barce, based on comments by Ned Barker, says that the natives were friendly to whites—"that true to their custom they were ready to afford assistance where assistance [was] required." Both whites and Potawatomis used canoes to hunt geese, swans, and ducks.[29] They all seem to have lived in the area without conflict, perhaps because the marsh and the lake offered an abundance for everybody. Campion points out that although there was friction between the Potawatomis and whites in the northeastern part of the state, relations between the two in the marsh area were peaceful.[30] The Potawatomis sometimes found living near whites very lucrative. A trader whose store was near the Kankakee reported paying three Potawatomi men over a thousand dollars for raccoon skins in one season.[31]

Settlers saw hundreds of Potawatomis in winter camp in the marsh in the late 1830s.³² These winter camps were important. According to white settlers who lived in the marsh,

> Indian Island near the Kankakee was the Indians' camping ground. During the hunting season of the fur-bearing animals, from early autumn to late in the spring, this island was their home. Along about the first of May they would pack their hides, furs, and with their squaws and children they would start for the Lake Michigan region to meet the fur traders.... The Indians would stay along the lakefront all summer, and early in the fall they would return to the camping grounds in the Kankakee swamps.³³

Settlers farmed in the marsh as early as the 1830s, perhaps before, harvesting marsh grass for hay and growing subsistence crops.³⁴ Hunters, both native and white, had long roamed the area, but beginning in the late 1870s, several hunting clubs were established. Such clubs as the Chicago Sportsman Club, the Rensselaer, the Diana Club, The Columbian Hunting Club, and others attracted sportsmen from all over the country.³⁵ They came both to hunt and to fish, for the marsh was famous for its huge flocks of waterfowl and ample bass, crappie, and pike. Meyer lists some of the names that whites bestowed upon the topographical features of the marsh, such as Indian Garden, French Island, Skunk Knobs, Bogus Island, Goose Island, and Long Ridge; and he adds the names of water forms, such as Flag Pond, Wild Cat Swamp, Devil's Race Track, and Frenchman Slough. Such naming testifies to an intimate knowledge of the marsh among white settlers and hunters. As Meyer points out, these landmarks identified homes, clubs, and campsites, and were also guideposts to the hunter traveling in a marsh where there were no roads.³⁶

All this activity, native and white, clearly shows a diverse and vibrant human presence in the marsh and the Beaver Lake area before they were drained, a presence that suggests a special cultural attitude toward swampy areas that is in sharp distinction to the view shared by many nineteenth-century Americans, most of whom agreed that wetlands should be avoided or eliminated. Michael Urban, in his account of Illinois wet prairie settlement, points out that early settlers saw wetlands as "evil landscapes," and quotes John Bunyan's seventeenth-century description as a good example of the fear and hatred expressed toward them: "This miry slough is such a place as cannot be mended; it is the descent wither the scum and filth that attends conviction of sin doth continually run, and therefore it is called the Slough of Despond."³⁷

Closer to home, in his 1935 study of Indiana wetlands, Richard Lyle Power comments that a "condition of forbidding wetness" stopped many, not only from settling, but even from traveling in northern Indiana.³⁸ Many Americans would have nodded in agreement reading this account of wetlands around Cairo, Illinois, written by Charles Dickens, traveling by steamboat on the Ohio River in 1842:

A dismal swamp, on which the half-built houses rot away; cleared here and there for the space of a few yards; and teeming then, with rank unwholesome vegetation... the hateful Mississippi circling and eddying before it, and turning off upon its southern course a slimy monster hideous to behold; a hotbed of disease, an ugly sepulcher, a grave uncheered by any gleam of promise, a place without one single quality, in earth or water, to commend it: such is the dismal Cairo.[39]

Dickens, no doubt, exaggerates. According to Hugh Prince, who quotes him at length in his study of American wetlands, Dickens wanted to warn prospective buyers about the dangers of buying property unseen. He was right to do so: speculators often bought tracts of swampland cheaply, then sold them to unsuspecting newcomers who had no idea they were buying land that was underwater. But hatred of wetlands was real. Commenting on legislation for the removal of wetlands, the *Congressional Globe* in 1850 encapsulated the prevailing attitude:

[Swamps and overflowed lands] are evils common to all countries, rendering, in their original condition, portions of the earth not only desolate and unsusceptible of cultivation, but fruitful promoters of disease and death. They can only be removed, or their evils mitigated by means of labor and money, which, when properly employed must redeem portions of the land from sterility, and make it valuable and useful, instead of the generator of disease.[40]

The *Globe* is referring to the passage of the Swamp Land Acts of 1850, possibly the most consequential environmental legislation ever passed by the US Congress. For years the states bordering the Mississippi River had wanted legislation that would grant unsold federal wetlands to the states for reclamation. In 1849 Congress granted to Louisiana the right to sell and drain its wetlands. On September 28, 1850, it made the same provision for other states with "swamp and overflowed lands," and stipulated "that the proceeds of said lands ... shall be applied exclusively, as far as necessary, to the purpose of reclaiming said lands by means of drains and levees." The Act gave over seventeen million acres of wetlands to Midwestern states alone; Indiana received over 1,254,000 acres.[41] Most of these were in the northern part of the state, and included the Beaver Lake lands and the Kankakee Marsh.

Indiana quickly passed its own legislation in 1851 and 1852, authorizing the draining of wetlands throughout the state for agriculture. Both McCorvie and Lant, and Hugh Prince, in their invaluable studies of wetland drainage in the Midwest, claim the Swamp Land Act was a failure. Prince argues that it "accomplished little or no improvement,"[42] and McCorvie and Lant say that it "contributed little directly to drainage of wetlands."[43] Both studies cite widespread fraud and incompetence on the part of state governments for this failure, and Indiana certainly had its fair share of both. (This is explored in chapter 2.) Nevertheless, the withdrawal from the public domain of millions of acres of wetlands allowed the states carte blanche in their disposal. By the end of the twentieth century, after 150 years of refinements to legislation, which strengthened local control of draining, and the introduction of new drainage technologies such as drain tiles, some 71 percent of Midwestern wetlands were gone.[44] The Swamp Land Act of 1850 was the historical incentive for this huge reduction.

The federal donation of the Kankakee Marsh to Indiana sealed its fate. Whether or not Beaver Lake had ever been included in the donation of 1850 was disputed (as discussed in chapter 2), but the first ditch to drain the lake was dug by hand in 1854–55 and paid for by Indiana. Thirty to forty years were required to transform its sixteen thousand acres into pasture and farmland. The first major plan to drain the marsh in 1869 failed (this is explored in chapter 3), delaying large-scale drainage until much later. Projects to dredge and straighten the river began in 1901; within twenty-two years, the marsh was gone, thanks to the industrial innovation of the steam dredge boat.

Indiana, despite setbacks to reclaiming the marsh, was relentless in pursuing its goal of creating new, lucrative farmland, even though the state already had abundant land under cultivation. By the end of the twentieth century, Indiana had lost 97 percent of its wetlands. In the light of current knowledge about the environmental value of such areas (see chapters 5 and 6), this doesn't seem like an accomplishment so much as what McCorvie and Lant aptly call a "massive environmental insult."[45]

Beaver Lake was drained before the projects to drain the marsh were begun, and its destruction, though initially paid for by the state, was completed by private landowners. Draining the marsh was also accomplished by private landowners. The state of Indiana entered not only the lake and marsh, but also the Kankakee River, into the domain of private ownership, specifically allowing landowners the right to modify the river in whatever way was necessary to drain the marsh. The success of the projects to convert the lake and marsh to cropland is a powerful testament to the imperatives of private property, enshrined in custom and statute, to preempt the interests of the public in maintaining control over land and waterways that were part of a shared commons. Draining the Kankakee Marsh was expensive. Many landowners strenuously protested against draining because of their unwillingness to be assessed to pay for it (discussed in chapters 3 and 5). There were no conservationist protests against draining either Beaver Lake or the Great Marsh in Indiana until well after these projects were finished (discussed in chapter 6).

Draining the marsh converted nearly half a million acres to cropland. Such a massive geomorphological change cannot happen without severe environmental consequences. Once major drainage projects began on the Kankakee River, it took Indiana less than twenty-five years to destroy the marsh. A hundred years later, Indiana and Illinois are still struggling to manage the consequences of that destruction (discussed in part 2).

# NOTES

1. Fay Folsom Nichols, *The Kankakee: Chronicle of an Indiana River and its Fabled Marshes* (Theo. Gaus' Sons, 1965), 193.
2. J. Lorenzo Werich, *Pioneer Hunters of the Kankakee* (n.p., 1920), 188.
3. Earl H. Reed, *Tales of a Vanishing River* (John Lane, 1920), 27.
4. Reed, *Tales of a Vanishing River*, 7.
5. Joel Greenberg, *A Natural History of the Chicago Region* (University of Chicago Press, 2002), 216.
6. Greenberg, *A Natural History*, 217.
7. Alfred H. Meyer, "The Kankakee 'Marsh' of Northern Indiana and Illinois" (PhD diss., University of Michigan, 1936), Papers of the Michigan Academy of Science, Arts and Letters, vol. 21, 362.
8. Greenberg, *A Natural History*, 218.
9. Meyer, "The Kankakee 'Marsh,'" 12.
10. Meyer, "The Kankakee 'Marsh,'" 13.
11. Greenberg, *Of Prairies, Woods, & Water: Two Centuries of Chicago Nature Writing* (University of Chicago Press, 2008), 84.
12. Greenberg, *Of Prairies*, 85.
13. Bartlett, *Tales of Kankakee Land* (Scribner, 1907), 1.
14. Bartlett, *Tales of Kankakee Land*, 24.
15. Timothy Edward Howard, *A History of St. Joseph County, Indiana*, vol. 2 (Lewis Publishing, 1907), 666.
16. Joseph McCarthy, *A Standard History of Starke County, Indiana*, vol. 1 (Lewis Publishing, 1915), 13.
17. *History of LaPorte County, Indiana* (Chicago, 1880), 343.
18. Rev. E. D. Daniels, *A Twentieth Century History and Biographical Record of LaPorte County, Indiana* (Lewis Publishing, 1904), 1.
19. *History of Porter County, Indiana*, vol 1 (Lewis Publishing, 1912), 11.
20. Louis Hamilton and William Darroch, eds., *A Standard History of Jasper and Newton Counties, Indiana* (Lewis Publishing, 1916), 221–23.
21. William Frederick Howat, *A Standard History of Lake County, Indiana and the Calumet Region*, vol. 1 (Lewis Publishing, 1915), 9.
22. John Ade, *Newton County, 1853–1911* (Bobbs-Merrill, 1911), 43–45.
23. Hamilton and Darroch, eds., *A Standard History of Jasper and Newton Counties, Indiana*, 243.
24. Burt Burroughs, *Tales of an "Old Border Town" and along the Kankakee* (Benton Review Shop, 1925), 108–10.
25. Burroughs, *Tales of an "Old Border Town"*, 111–12.

26. Elmore Barce, *Beaver Lake, A Land of Enchantment* (The Kentland Democrat, 1938), 70.
27. Barce, *Beaver Lake*, 56–57.
28. Charles Test, "Beaver Lake: a Reminiscence," *Indianapolis Journal*, November 11, 1877.
29. Barce, *Beaver Lake*, 91.
30. Thomas Campion, "Indian Removal and the Transformation of Northern Indiana," *Indiana Magazine of History* 107, no. 1 (March 2011): 56.
31. Campion, "Indian Removal," 56.
32. Campion, "Indian Removal," 56.
33. Meyer, "The Kankakee 'Marsh' in Illinois and Indiana," 8.
34. Meyer, "The Kankakee 'Marsh,'" 10.
35. Meyer, "The Kankakee 'Marsh,'" 11–12.
36. Meyer, "The Kankakee 'Marsh,'" 13.
37. Michael Urban, "An Uninhabited Waste: Transforming the Grand Prairie in Nineteenth Century Illinois," *Journal of Historical Geography* 31, no. 4 (2005): 651.
38. Qtd. in Hugh Prince, *Wetlands of the American Midwest, A Historical Geography of Changing Attitudes* (University of Chicago Press, 1997), 119.
39. Qtd. in Hugh Prince, *Wetlands*, 121.
40. Qtd. in Mary McCorvie and Christopher Lant, "Drainage District Formation and the Loss of American Wetlands, 1850–1930," *Agricultural History* (Autumn 1993): 24.
41. Prince, *Wetlands*, 145.
42. Prince, *Wetlands*, 205.
43. McCorvie and Lant, "Drainage District Formation," 28.
44. McCorvie and Lant, "Drainage District Formation," 22.
45. McCorvie and Lant, "Drainage District Formation," 22.

# 2

# SELLING BEAVER LAKE, 1853–1891

Fraudulent Ditches, a Disappearing
Lake, and a History of Absence

DRIVING SOUTH ON US 41 FROM LAKE COUNTY, INDIANA, INTO NEWton County, a traveler might notice a change in scenery after crossing the Kankakee River, which marks the boundary between the two counties. Six or seven miles south of the river, the pleasant view of cultivated farmland bordered by woodland changes, and the traveler seems to be driving through apparently uncultivated prairie. The land stretching away from the road both east and west seems colorless, featureless, and has few visible landmarks. From here to Morocco, another seven miles or so, the road is almost ruler straight, and the scenery changes little. An observant traveler might notice a sign on the side of the highway that suggests a reason for the sudden change in landscape. About seven and a half miles south of the river, the sign stands within a small parking area on the west side of the road and is worth stopping for. It explains that the visitor has entered the basin of an ancient lake, and the view as far as the eye can see in any direction marks the grave of that lake. Information about the lake's size is offered, along with selected historical information, including the fact that the lake was the largest lake in Indiana until it was drained. Information about when it was drained, why, or by whom, is not included.

The sign was erected by the Newton County Historical Society in 2010 in observance of the county's sesquicentennial anniversary. It shows the outline of the lake as it appeared on nineteenth-century maps, and current road maps of Newton County include this outline. The maps show a lake whose borders were in three townships, filling almost all of McLellan Township from the south into Lake Township in the north, and west from nearly the state line east into Colfax Township. The sign and the maps suggest something about the values of the citizens who created them, citizens who want the lake to be remembered, and so have memorialized it, not only with a sign, but also as a ghostly reminder on their maps.

The old lake bed now contains farms, residences, roads (US 41 runs through the heart of it), and a school—it has become a geography of what historians call the "built environment." John Jakle advocates for a reading of Indiana's built environments as "rich texts" where "[p]astness in landscape is a visible accumulation of overlapping traces from successive periods."[1] Nothing in the landscape around the sign on US 41 shows a visible trace of the period when the land was covered by a lake. Perhaps a true history of place should also include a history of *absence*. This would be a history of what was *dis*placed in order to create a space for farms and roads. Beaver Lake as an actual geographical entity is not only extinguished—all evidence of its existence has been obliterated. When one stands in front of the sign commemorating it, one stands on ground once covered by water—thirty-six square miles of water—a place whose look, and indeed whose entire geographical and biological character, has changed utterly. The sign is a lonely statement about absence. It wants us to know that we are standing on ground where something vital is not only missing, but perhaps, where something vital was *taken*. If not for the sign, who would know?

Some believed that it was, indeed, taken. Writing in 1877, Charles Test, who was the first circuit court judge in Newton County, blamed "the vandal hand of the speculator" for the lake's destruction. He goes further, outlining the strategy used: "They bought the land contiguous to its borders, then drained off its waters to the Kankakee by deep ditches, and claimed its bed as riparian owners." Charles Test probably knew something about what he was talking about, since he was personally acquainted with the men he criticized.[2] (Test is a paradoxical figure in this narrative, as discussed later in this chapter. He both defended—and condemned—the men he refers to here.) The historical record shows that the judge was correct, but understated the case: speculators almost certainly had the lake drained, probably illegally, but they could not have succeeded without the help of state officials. An examination of events surrounding the draining of the lake, and the long controversy that centered on the ownership of the Beaver Lake lands, reveals a decades-long history of fraud, ineptitude, and greed. This history involves landowners, officials at the county and state level, the Indiana government, the US Land Office, and the US Congress.

Since the saga of Beaver Lake is a story about ownership, it is proper to begin with the initial survey of the Beaver Lake lands. As territories became states, the US Land Office sent out teams to conduct a rectangular survey of the new federal acquisitions. This survey consisted of a grid system that marked off townships of thirty-six square miles in each county. Townships were subdivided into sections, quarter sections, lots, and so on. When surveys were complete, land was sold, and entries to properties were registered in local land offices.[3] The grid system was essentially a way of commodifying land, of creating "parcels" whose precise locations and dimensions could be measured and recorded. Most of the land in new states was public domain, and it had to be surveyed in order to be sold. This fact will become important later in our story.

Surveyors were the reconnaissance teams of white settlement. They were instructed to note any features that might be useful to potential settlers—such as waterways, swamps, kinds and amount of timber, the quality of soil, mineral, salt licks, mill seats, etc.—and they were expected to provide detailed outlines of lakes.[4] Surveying could be dangerous work. Tribes in some states knew what the teams were up to, and murdered them routinely.[5]

Surveyors came to Beaver Lake in 1834 and finished their work in 1835. Luckily for Samuel Goodnow, deputy US surveyor for Jasper County and his team, the wigwams they saw around Beaver Lake in January 1835 were inhabited by friendly Potawatomis. The surveyors must have interacted with them because Goodnow wrote in his field notes that the "the lake was named 'Devil's Lake' by the Indians." Goodnow saw the Beaver Lake country as useless for settlement. He wrote that the township containing the lake had "no good land, very little good timber, no springs or stone quarries have been discovered [and] it is considered as being of very little value." He wrote, "[t]his township is all a lake, or deep marsh and morass, except a little in Southwest corner. Marsh 4 or 5 ft. deep. No outlet to lake as I discover. Timber is scattering and scrubby." Goodnow and his team did provide an outline of the lake. They meandered its borders, and this became the iconic outline of Beaver Lake that appears on present-day county maps.[6]

In 1851, an observer saw the lake as "a clear smooth sheet of water, about seven and a half miles long and five miles wide, and the water was from two to nine feet in depth."[7] Estimates of its total acreage varied from fourteen thousand to sixteen thousand acres, but accounts agree that its margin was fringed by marshes of varying depth and width.[8] Charles Test's description of the lake and surrounding area was typical, though his opinion of its beauty was not. Reminiscing about a visit in 1857, he wrote that the lake was

> a beautiful expanse of water covering nearly a township of land... to approach from the south or east was difficult owing to the extensive marshes and quicksands which laid in the way... It was indeed a wild spot. Not a human habitation located within miles of its border. The lake itself abounded with fish and wild fowl; also deer, wolves and muskrats in considerable numbers.[9]

In the 1850s most people in Indiana—indeed, most Americans—would not have shared Test's opinion of the lake as something rare and valuable in itself. Most would have agreed with the surveyor in 1835 that it was worthless. As mentioned in chapter 1, in 1850 the Swamp Lands Act granted 1,265,000 acres of wetlands to Indiana. Included in this grant were fewer than nine thousand acres in the township where Beaver Lake was mostly located.[10] This number probably reflects the acres of marshland around the lake. It did not include the lake itself, which was approximately sixteen thousand acres. This fact is important because, as will be made clear later, the federal government never considered Beaver Lake as "swamp and overflowed lands," as defined by the language of the Swamp Lands Act. In other words, as far as the federal government was concerned, the lake itself was never intended to be drained.

Indiana was eager to comply with the provisions of the Swamp Lands Act. In 1851 the General Assembly pledged quickly to pay down the public debt with revenue from sale of the land.[11] Wetlands occurred in nearly half of Indiana's counties, but the largest concentrations were in the northern part of the state. This included Jasper County, where Beaver Lake was located. (Newton County, containing Beaver Lake now, was formed from Jasper in 1860.) The Winemac District, which comprised thirteen northern Indiana counties, contained over eight hundred thousand acres of wetlands, well over half the total federal grant to the state.[12] This area was not well developed because of the prevalence of boggy land—settlers avoided it because cultivation was extremely difficult, and disease was common.[13] The state now saw a golden opportunity to make the northern part of the state fit for settlement. The General Assembly believed that "a desolate waste" would be transformed "into a habitat for an industrious, happy and healthy people."[14]

From 1851 to 1852 the Assembly passed its own swamp land legislation and worked out a procedure to administer the sale and drainage of wetland areas. Sale of land was to be through public auction, and the state pledged to drain the land with the proceeds. In 1851 a new state constitution mandated that proceeds from wetland sales would go to the public school fund, minus the cost of draining. In 1853, auctions for the sale of wetlands were held in the counties where they were located, and revenues were good enough to encourage the next step: reclamation. The process for draining began with a county swamp land commissioner, appointed by the governor. The commissioner was to consult with an engineer, accept bids, then proceed with drainage work at the lowest cost.[15]

By 1853, the governor had appointed the first swamp land commissioner in Jasper County, John Darroch, who began the process of holding auctions for ditching by advertising in local papers. Darroch held such an auction, probably the first in Jasper County, in November 1853; he notified contractors of it in October of that year.[16] Darroch, together with County Engineer James Ballard, was responsible for designating the location of the ditches. Usually, "Notice[s] to Contractors" specified the exact location of where ditches were to be dug, as in this one, published in February 1854. (Interestingly, a notice for Beaver Lake Ditch was never advertised, as discussed later in the chapter.)

Ditch No. 1. Commencing at Station 1[?]4, about 330 feet below the Bridge on the road from Renssalear to Lafayette, in Big Slough on Section 12, Town 28 N. R 7. W. running East from the nee [*sic*] along the channel of said Slough to the County line about 10 miles in length.

Ditch No. 8, Commencing 700 feet E. of S.W. qr. of S. 12, T. 31 N.R. 10. W. and running N. 14 deg. W. to Station 22 then N. 31 deg. W. to State line 1-2 mile in length.

Ditch No. 11. Commencing at Station 263, of Ditch No. 1 on Section 9. Town 23, N.R. 6 W. and runs S.E. and S.W. through Section 9, 10, 15, 16, 21, 22, 28, 29, and 30, of the above Townships, and about 5 1-2 miles in length.[17]

The only restriction the state placed on ditching contracts was that they were to be given to the lowest bidder. The county swamp land commissioner and county engineer had wide discretion as to where ditches were to be located, and the swamp land commissioner had sole authority to commission the construction of a ditch. There was no requirement about ditching *only* to drain "swamp land." The legislature did not regulate ditching beyond specifying the forms that swamp land commissioners had to use to certify the contracting for ditching. Though various methods of draining were used in Indiana and other states later in the nineteenth century, in the 1850s, the primary method was to dig ditches by hand.[18] Extremely labor-intensive as it was, ditching was regarded by the legislature as "producing surprising and valuable results." This "system of open ditching" was seen as well-thought-out, comprehensive, and efficient:

> The ditches are varied to suit the amount of water to be discharged. If the quantity be large, the ditch is made wide and open, if small, it is more narrow and shallow. By making the main outlets quite deep, sufficient fall is usually obtained to answer the desired purpose of discharging the surplus water. Several main branches are generally made to lead into one of these outlets, and multitudes of small drains into them, thus forming a system connected with each other like a river and its tributaries.... The deep ditches are usually preferred on black soils and shallow ones in clay soils.... The deeper the ditch the greater the results, provided the outlet be sufficient.[19]

This was the system intended by the General Assembly to drain the wetlands of Indiana and transform them into productive farmland. It is certainly the system that eventually drained Beaver Lake. For several years, however, the state paid a great deal of money for ditches that were never constructed, or that were useless for effective draining. Fraud was rampant, especially in the northern counties. This scandal has been well-documented by Stephen Strausberg, who points out how vulnerable the system was at the county level:

> By the fall of 1857 the entire program tottered on the verge of collapse. The state auditor, John Dodd, accused the swamp lands commissioners of allowing the tracts to fall into the hands of speculators who had no intention of fulfilling the terms of the agreement. He suggested that each contract should contain a provision rendering it void if the work was not satisfactorily completed within a reasonable time or by a specified date.[20]

In 1859 and 1861, the General Assembly mounted investigations into seven counties, including Jasper, which examiners singled out as the location of one of the worst scandals. "In the county of Jasper," they commented, "our investigations have satisfied us that the officers of that county have not only aided others in the commissions of great frauds upon the swamp land funds, but have also been participants in the profits arising therefrom."[21] Between 1853 and 1858, thirty individuals and partnerships were awarded over ninety contracts in Jasper County for dozens of ditches at a total cost of over $187,000.[22] One man—Swamp Commissioner John Darroch's father-in-law, A. M. Puett—was paid nearly $37,000 on several contracts, far more than anyone else.[23] In 1859, the Indiana Senate investigating committee reported that "not one half of the ditching [in Jasper County] is yet completed."[24] Few records of these transactions could be found, and no one in Jasper or in any other county was prosecuted, although a few individuals were sued.[25] Of the $1,759,752-worth of wetlands that Indiana sold in total, $1,674,932 was spent on drainage programs and administrative expenses.[26] That means that Indiana spent over 95 percent of swamp land revenues on drainage with little actual productive farmland to show for it by 1860, and very little money left over for the school fund.

One of the few projects that was completed in Jasper County was given to Puett, begun in 1854 and finished in 1855.[27] This was the first ditch excavated to drain Beaver Lake, and became known as the Beaver Lake Ditch. It connected the northern edge of the lake with the Kankakee River, a distance of about four and a half miles.[28] Circumstances surrounding the location of this ditch are suspicious. The ditch does not appear on the list of ditches submitted by the county swamp commissioner, John Darroch, to the Senate investigating committee in 1859 (more about this investigation later). No plans or specifications for the ditch were submitted for public view, and the ditch was apparently never bid on at public auction.[29]

This site may have been chosen because it seemed the most likely to drain the lake quickly. Amzie Condit, state engineer for Northern Indiana, along with others, had surveyed the terrain between the lake and the Kankakee River and found a "fall" of forty-two feet between lake and river. According to Condit, when the ditch was complete and the channel opened, "water rushed out with great force and noise, tearing out huge trees by the roots."[30] As a college-trained civil engineer, Condit would have known that a large volume of water would be released by a ditch cut through such an elevation.[31]

As early as summer 1853, plans to drain Beaver Lake had been publicly discussed by state officials.[32] In November of that year, months before ditching began, Condit and the state Auditor, John Dunn, purchased the entire shoreline of the lake, some nineteen miles by Condit's estimate.[33] A month later, they sold almost all this land to Michael Bright. Bright, like Condit and Dunn, was among the many land speculators who descended upon Jasper County in 1853 and 1854, and given the level of corruption in that county, the press was quick to suspect an illegal scheme.[34] In April 1854, a reporter for the *Indianapolis Journal* offered an opinion as to how Bright would eventually own the entire lake bed:

> [Bright] now owns, I learn, the entire body of land bordering on Beaver Lake, which covers 17,000 acres. This lake is to be drained, and Mr. Bright will, it is said, become the owner of all the land submerged by the lake. If an act of Congress should be necessary to perfect this title, why of course Jesse D. can procure that.[35]

No act of Congress was necessary to "perfect" Bright's title, as it turned out. But this suggestion underscores the writer's suspicion that Bright's desire to own the lake bed was legally dubious. "Jesse D." was Jesse D. Bright, brother to Michael, elected three times as a US senator from Indiana. Jesse Bright was a powerful political figure in the state and boss of the Democratic Party for many years, when the Democrats controlled most offices in Indiana.[36] (Governors of Indiana were all Democrats from 1843–61.) Michael G. Bright was a wealthy lawyer from Madison, Indiana, whose dealings throughout the state were closely watched by the press.[37] The writer for the *Indianapolis Journal* didn't quite have it right in 1854: Bright had not yet become the owner of *all* the land bordering the lake. But he was certainly correct that Bright's ambition was to own all the land *under* the lake. Michael Bright is a pivotal figure in the story of the destruction of Beaver Lake. His actions were to entangle the issue of ownership of the Beaver Lake lands in a controversy that persisted for over thirty years.

In 1857, Michael Bright paid $10 to Dunn for the last few lots of the lake shore that he had not bought in 1854.[38] He now owned every fractional lot around the lake, and he set about solidifying his claim to the entire lake bed. His method for accomplishing this was to declare that his ownership of the lake shore conveyed "riparian rights," which entitled him to ownership of the entire lake bed. This novel claim is described in *The Standard History of Newton and Jasper Counties*:

[Bright] made a plat of these marginal lands and of the entire lake. In this plat he assumed to extend into the lake the outward lines of the Government survey, east and west and north and south, so, as he asserted they would, by due intersection sub-divide the entire area into lots of forty acres each, which lots he numbered on this plat from 1 to 427. Attached to this plat he made a written statement, which ... contains this declaration of ownership: "And whereas, in virtue of being riparian proprietor of all of said lots and tracts of lands, I am, by operation of law, the owner and proprietor of the bed of said lake, and of all the islands covered by the waters thereof."[39]

The "operation of law" Bright referred to was, at best, ambiguous—more likely, completely false. His claim would eventually be repudiated (explained later), but his "deed" was good enough for the recorder of Jasper County (a county soon to be notorious throughout the state for land fraud), and was duly entered. In 1857 almost all of this land was still underwater, but the draining of the lake was underway, and Bright wanted to make certain that there should be no hindrance to future attempts to complete the draining begun three years before. He added this to the deed for Beaver Lake:

[T]here is hereby reserved to the present proprietor and to any party or parties who may hereafter be and become the owner of any of said lands, the right at any and all times, of opening and keeping open the ditches or drains which have been undertaken for the purpose of draining off the waters from said Lake, and to this end they or any of them may at any time enter upon so much of said lands as may be necessary for the purpose and may repair and open said ditches, and if essential to the further reclamation of any of said lands may deepen and enlarge these so as to secure the largest possible drainage, doing to the other lots and lands as little damage as may be consistent with the accomplishment of that object.[40]

Bright knew that current ditches needed to be maintained, and that more would be required. He may have claimed all of Beaver Lake as his private property, but the state continued to bear the cost of draining. In spite of the initial spectacular success of Beaver Lake Ditch, only about two and a half feet of water had been drained off by 1859. The ditch was shallow and overflowed; it had to be cleaned out and recut at a cost of $600. Also in that year, a contract was let to widen and extend the ditch about a mile into the lake. However, the lake continued to refill from rainfall.[41] Ditches required continual repair, and more would be needed.

Bright knew that his title to Beaver Lake, even though entered into county records, was legally shaky. His "riparian" deed manufactured an impossible survey of land that was underwater—it was essentially a forgery, since no US surveyor had actually conducted it. Bright realized that his claim to ownership of the entire lake bed might not stand close scrutiny, so he devised a way to give his sham title some semblance of legality. In 1857 he brought suit against William Blake, who, Bright claimed, had illegally taken possession of an island in the lake. The Jasper County Circuit Court upheld Bright's claim to the island as part of his self-proclaimed riparian title to the lake bed. However, it was widely believed that Bright had paid Blake to make this claim so that Bright could sue, and win.[42] As the *Kentland Gazette* put it, "[t]he purpose of this miserable sham suit was to strengthen the claim of Bright to the bed of the lake, and to create public confidence in it so that he could negotiate its sale."[43] As it turned out, Bright was able to leverage not only the circuit court, but also the Indiana state legislature into endorsing his claim.

Michael Bright himself had been a member of the state legislature, was a delegate to the 1850 constitutional convention, and had been agent of the state.[44] He was well-known among those in power, and was deeply entangled in the business affairs of the state and with state officials. Bright's affairs with the state were complex, and his land holdings, including those around Beaver Lake, played a key role in these transactions. Bright was partner with John Dunn, state auditor, in a suit brought by the Ohio Life Insurance and Trust company against Dunn in 1856. A judgment was levied against Dunn, and to pay it, Bright offered 795 acres of Beaver Lake shoreline. Bright's brother George bought the land in November 1857, and the money was used to pay the judgment. Michael Bright was intent on maintaining his claim to the entire lakeshore, however, and that same month, George Bright sold all of this land back to Michael for $100.[45]

In 1855, W. R. Nofsinger, state treasurer, loaned $40,000 in state money to Allen May, a partner of Bright's. When the debt came due a few years later and May could not pay, the state sued Nofsinger and May, as well as Bright, who had pledged surety to May. Bright offered lands in Jasper County as partial payment for May's debt, as well as 1,280 acres near Beaver Lake. Nofsinger eventually settled with the state for the remainder of the debt.[46]

In 1858 and 1860 Bright concluded transactions that would convey to the state one half of his riparian claim to the bed of Beaver Lake: 7,880 acres of land. The circumstances surrounding this transaction are far from clear, but Newton County records show that Bright sold the lands to Aquilla Jones, state treasurer, in 1858, and Jones conveyed these same lands to the state in 1860.[47] According to the Auditor's Report of 1860, the legislative commission accepting these lands had been very careful about confirming the title of other lands taken from Bright; however, they accepted nearly eight thousand acres on a dubious title without flinching. The US Congress would eventually repudiate Bright's riparian claim, and the Indiana Supreme Court would invalidate it years later, but in 1860, the Indiana legislature effectively legitimized his claim to ownership of nearly sixteen thousand acres of land that he had never paid for, that had never been surveyed, and that were mostly underwater. The only title Bright had to this land was of his own manufacture, and the state gave that title its seal of approval.

In 1862, Bright began selling his half of the lake; in 1865 the state reconfirmed his title by putting its share of the Beaver Lake lands up for sale. Bright knew that questions might arise about his right to sell these lands, so was careful to disclaim any "defect" in his title: "Said parties of the first part hereby covenanting and agreeing that they have done no act to impair or lessen the title to said premises, but are not to be responsible for any acts or omissions of others or for any defect whatever of title."[48]

In spite of Bright's admission that his title could be questionable, by 1871 most of the lake bed had been sold. Algy Dean, Lemuel Milk, and another Milk partner, his nephew Henry Cooley, all wealthy landowners from Illinois, bought the greater part of it, owning among them over twelve thousand acres, some three-quarters of the lake bed.

Sometime after 1861, Algy Dean and others extended Beaver Lake Ditch another two miles into the lake, and other ditches were added. By 1871 the Beaver Lake Ditch was almost seven miles long, from forty-five to seventy feet wide, and from two to twenty-four feet deep. A contract had been let to extend the ditch even farther.[49] A lake that fewer than twenty years before had covered an entire township and more—an area greater than thirty-six square miles—was now reduced "to a section and a half of water... about three miles in length and about one-half mile in width." Most of the water left was not over twelve inches deep.[50]

By 1870, twenty houses had been built on the lake bed. The population of McClellan Township, where Beaver Lake was located, was 141.[51] Much of the land was under cultivation, and most had been fenced. Many of those who farmed there were probably tenants of Milk or lived on land mortgaged to him. Milk and Dean were extremely protective of their interests in the Beaver Lake land. Dean was known to threaten those he considered squatters, force them from their homes, and destroy their crops.[52]

By all rights, the story of Beaver Lake's destruction should end here. Michael Bright had stolen the lake from the public domain, made a small fortune selling land he never paid for, and Lemuel Milk and his associates were well on their way to erasing all traces of the lake. But in the spring of 1871, something extraordinary happened. A group of men arrived at Beaver Lake who were determined to challenge Milk's and others' rights to the old lake bed. In fact, their object was to disprove the validity of the Bright title, upon which Milk's titles rested, and claim the Beaver Lake land for themselves. Their claim was to spark two investigations by the US Congress, require Congressional legislation, and cause legal fights in the Indiana courts and in its legislature that were to last nearly twenty years.

These spring arrivals, about forty men led by Isaac Marker Dresser and Amzie Condit (the very same Amzie Condit who had originally bought the lake shore and helped design the Beaver Lake Ditch in 1854), showed up at Beaver Lake in May 1871. The most colorful account of what happened next is from the *Chicago Tribune*, reprinted in the *Cambridge City Tribune* in August 1871:

> The party started out for Beaver Lake equipped with some 50,000 feet of lumber, wagons, cattle and tools, and all the implements of a party of pre-emptors. On arriving at the lake, they were hospitably met by a company of about 100 persons, who were mounted and on foot, bristling with guns, rifles and other lethal weapons, under the order of Mr. Algea [*sic*] Dean, a farmer in the service of Mr. Lemuel Milk. The first courteous greeting with which they were startled came from Mr. Dean, to the following purport: "We are come to notify you Chicago roughs and robbers, that yew must leave this yer neighborhood, or suffer the consequences by remaining." Mr. Myers declined to leave the neighborhood, and stated his intention of pre-empting the land. This they were not permitted to do, having been met at all points by armed men, who threatened demolition.[53]

This exact scenario is unlikely. Other, more credible, accounts say the pre-emptors, though probably threatened by armed men, had actually begun erecting wooden structures before they were discovered and taken to a local justice of the peace. A Chicago attorney argued their case, and they stayed on Beaver Lake at least throughout the summer while the matter was litigated.[54]

The pre-emptors insisted that the bed of Beaver Lake had never been surveyed, that it had never been included in the donation of swamplands given to the state, was therefore still the property of the United States, and thus available for pre-emption (ownership by right of first possession under the Preemption Act of 1841). They applied to the commissioner of the General Land Office for a survey so they could make suitable claims. The commissioner ruled that the Beaver Lake lands had not been conveyed to the state because the lake had never been considered swampland; the lands were therefore still the property of the federal government, and that the riparian title used by Bright to sell the land to the state of Indiana and to others was invalid.[55] However, what looked like an easy victory for the pre-emptors was quickly deflated. Lemuel Milk had a lot of influence. He appealed to the governor of Indiana to intercede with the commissioner, and he did. The commissioner agreed to suspend the survey until the matter could be settled by the US Congress.[56]

The merits of the pre-emptors' case inspired vigorous debate in the Indiana press. At the core of the issue was the legitimacy of the Bright title to the lake bed, for if his title was invalid, every sale of Beaver Lake land was invalid. This brought into question the integrity of the state of Indiana, which had bought half of the lake bed from Bright and sold it to Milk, Dean, and others. Anonymous writer "X" published a defense of the state's claim in *The Indianapolis Journal*:

Parties in buying from the State supposed the title good, and relied upon her deed, and in good faith ought she not attempt to maintain the title? It is a sentiment of common honesty that one who sells lands, takes the money, and makes a deed, ought to do all that properly may be done to protect the right of the purchaser... Common honesty demands that the State authorities should stand by those purchasing, and protect their titles.[57]

Two weeks later, another anonymous writer, using the name "Y," published a lengthy repudiation of X's claim about the "honesty" of Bright's, and the state's, title in the same newspaper:

The deed of conveyance made by Bright in this case is a curiosity in itself. It is patent on the face of it to the most obtuse intellect that Bright knew he was perpetrating a fraud, and had no confidence in the validity of the conveyance that he was making, and that the officers who took the conveyance for the state must have known that she was receiving a title that was wholly worthless, and that they must have been parties to the fraud with Bright.

"Y" believed that the scheme to claim the lake bed as a "riparian speculation" was the reason that the lake had been drained in the first place, and offered a surprisingly detailed history of the circumstances surrounding the draining of the lake:

A. B. Condit, the first Swamp Land Engineer appointed for this part of the State, ascertained that this lake was four and a half miles from the Kankakee River, that in that distance there was forty-two feet fall; that if the waters were started through the sand ridge to the river, they would, from the nature of the soil—sand and loam—cut out a channel sufficiently deep to drain off the entire waters of the lake, and that the draining of the lake would drain, or greatly facilitate the drainage of some 25,000 acres of swamp land around the vicinity of the lake. Mr. Condit, having heard of the doctrine that the riparian owner would own to the center as the waters receded, concluded it would be a fine speculation to own the lake rim. Not having the funds to purchase the entire rim, he associated with himself John P. Dunn, the Auditor of the State, which office enabled him to command free bank money ad libitum, and the fractions composing the rim of the lake were purchased by Condit and Dunn. M. G. Bright afterwards brought [sic] these fractions of Condit and Dunn and thus became the "sole proprietor" of the riparian speculation.[58]

"Y" was not the only one to publish an account of a conspiracy to drain the lake and sell off the lake bed. J. W. Conner, editor of the *Kentland Gazette* (in Newton County), wrote to expose facts that "speculators and free-booters" were attempting to hide:

Some years ago Michael G. Bright trusted a certain party with money enough to enter the surveyed fractions around Beaver Lake, and the fellow entered them in his own name, and then Bright had to pay this fellow a bonus to get the title. With the title to these, Bright set up his claim to the entire bed of the lake, by riparian right. About this time, Mr. John Darroch, of this county [Newton], being a personal and political friend of Bright, let the draining of the lake to his father-in-law, A. M. Puett, for $13,000.[59]

The US Congress demanded an accounting of what had happened to Beaver Lake. In December 1871, the US Senate passed a sweeping resolution regarding the dispute. Among other things, the resolution demanded an explanation for why the lake had been drained. It called for the secretary of the interior to report to the Senate when, by whom, at what cost, and under what authority the lake was drained, how much land had been drained, its value, what profits had been made from its occupancy, what legal proceedings had transpired to determine title to the lands, and what legislation might be necessary to enable the commissioner to adjust the competing claims to the lake.[60] A similar resolution was passed in the House, and in early 1872, the secretary of the interior presented reports to both the Senate and the House of Representatives.

The report to the Senate was the most thorough. Its sixty-six pages contain specific answers to all the questions posed in the December resolution, backed up with detailed written statements from Newton County surveyors. Statements were also gathered from a former swampland commissioner, assorted Newton County officials, and men who purchased property on the lake bed. There were also statements from Isaac Myers and Parker Dresser, leaders of the pre-emptors, refuting the Bright title and supporting the pre-emptors' right to the Beaver Lake lands. The governor of Indiana sent a statement but refused to take sides. The report also included a complete list of every sale of Beaver Lake lands, from the first state patent of lakeshore property sold in November 1853 to Condit and Dunn, to the last sale of titles based on Bright's riparian claim in October 1871. Most of these entries contain date and amount of sale, survey locations (township, range, section, and lot), and acres sold. It is possible, in other words, to trace every single sale of Beaver Lake lands for nearly twenty years and discover who profited from them. All descriptions and dates for purchases of this land in this article for that time period are based on this list.[61]

The prestigious Indianapolis law firm of Hendricks, Hord, and Hendricks represented Milk, Bright, and the state of Indiana. The firm wrote an extensive statement defending Bright's riparian title and arguing for Indiana's right to ownership of the lake bed.[62] Milk also hired Charles Test, former circuit judge of Newton County, to represent him. Test sidestepped the issue of Bright's riparian title and emphasized Indiana's right to own the lake bed. Test's role in the story of Beaver Lake is ambiguous at best, as mentioned earlier in this chapter. In his brief advocating for state ownership of the lake, he describes it as "a waste of waters.... useless for any commercial or other purposes."[63] This contradicts his moving description of the lake, quoted in chapter 1. He also says that the state drained the lake, which directly contradicts his statement, quoted earlier in this chapter, that speculators were responsible for its draining.

Test later defended his actions before the Senate by saying that he was never Bright's counsel, and that his only motive was in ensuring that Indiana should have title to Beaver Lake.[64] Test seemed to want it both ways: in 1877 he condemned the men he defended in 1872. He makes no mention in his 1877 reminiscence of his 1872 explanation of defending Milk, published in the same newspaper. We can only conjecture that Charles Test, writing as a lawyer, believed that his actual experience of Beaver Lake had no real bearing on the legal argument he was constructing. This doesn't explain his contradictory statements about who was to blame for draining the lake. Test knew the truth about that. Perhaps he felt that by blaming Indiana, he somehow bolstered the state's right to ownership.

The Secretary's report to the House was shorter—only ten pages—and consisted mostly of written depositions from some of the pre-emptors, including Amzie Condit. Michael Bright, whose riparian claim to the lake bed was the root cause of these investigations, did not issue any statement that is part of the historical record, nor did the purchaser of most of Bright's claim, Lemuel Milk. Algy Dean, however, used the occasion of a written deposition to threaten Amzie Condit and Isaac Myers, leaders of the pre-emptors:

> Myers Condett [sic] and others take notice that I forbid you entering the enclosure known as Milk's and Dean's Beaver Lake lands until our ownership is at an end. If you persist in doing so, I shall treat you as though you were entering my door-yard against my orders.[65]

Dean's and Milk's ownership was not at an end—far from it—even though the lawyers' arguments for Bright and Milk did not prevail: both houses of Congress agreed that Bright's claim was bogus, and that he had no legal right to sell the bed of Beaver Lake, either to the state or to anyone else. The title to Beaver Lake rested solely with the United States, and had never been relinquished. This meant that all other titles, including Milk's, were invalid.

Congress recognized a dilemma, however, born of two salient facts. One was that the lake, which was never intended by the federal government to be drained was, in fact, almost entirely drained, and the other was that individuals had purchased its land in good faith, and built houses and farms and roads. (The issue of "good faith" is subject to debate. See Congressman Townsend's remarks below.) In other words, the conditions under which Bright had perpetrated his original fraud had changed (the lake was now almost completely destroyed), and even though those conditions might have changed precisely because of that fraud, they could not be undone, and besides, in the eyes of almost everybody, the Beaver Lake lands had been "improved" with the removal of its wetlands, and with the construction of fences and roads.[66]

The Senate's answer to the dilemma was to sidestep the issue of fraud, and to give the problem back to Indiana. In February of 1872, Indiana senator Daniel Pratt introduced a bill that would grant title of all Beaver Lake lands to the state.[67] The Committee on Public Lands in the House of Representatives was charged with reviewing the Senate bill, and reporting to the House. The chair of that committee, Washington Townsend, warned against assigning title of the lands to the state:

> [I]f Congress should quit-claim to her, or make valid all the titles to all the reclaimed lands, such quit-claim would be a congressional recognition of the great land robbery of Bright and Dunn,[68] who, by the purchase of 2,000 acres or less on the margin of and encircling the lake, would thus be recognized as having properly seized on 15,000 acres more, worth from 850,000 to $100,000, without having paid one dollar for them.

Townsend understood that more than a matter of titles was at stake. Lands all over the country were in danger from speculators like Bright, and Townsend's committee had only a month before recommended passage of the bill that was to create the first US national park, Yellowstone.[69] He spoke for a growing awareness that land could have value beyond the monetary:

> [An act to quit-claim] would establish a dangerous precedent that would be quickly followed in this era of land speculation, and the nation would have some of its most beautiful lakes dried up, and be robbed of some of its most valuable territory for the benefit of land-grabbers in a manner of which this Beaver Lake land-grab is a conspicuous example.[70]

Townsend's committee recommended against returning the land to the state; rather, the federal government should sell it to current occupants for $1.25 an acre, the established price for government land. Townsend acknowledged that such an arrangement might be "hard" for those who purchased land from Bright, but he pointed out that "they should remember that Bright refused to warrant the title in his deeds, and they, therefore, had notice that he doubted its validity, and they cannot complain."[71] As far as Townsend was concerned, Bright had clearly labeled his titles "caveat emptor," thereby invalidating any notion of buying in "good faith" on the part of purchasers. They had been defrauded, but they had been told that they were being defrauded.

In the end, however, the Senate bill prevailed, and on January 11, 1873, the US Congress released title to the Beaver Lake lands to the state of Indiana.[72] Now it was up to the Indiana legislature to decide what to do with the fact that the state now officially owned the bed of Beaver Lake. The state had bought half of the Bright riparian claim in 1860 and had sold it in 1865. Now by an act of Congress it found itself the owner of the other half—the half sold by Bright to Milk, Dean, and others. The question for the state was not only how to recover this land, but whether or not it really wanted to. Indiana was squaring off against Lemuel Milk, one of the richest and most powerful men in the state. By some accounts Milk owned, or was to own eventually, sixty-five thousand acres in Indiana and Illinois, forty thousand acres of which were in Indiana.[73] Since Milk's holdings were centered in McClellan township, Newton County, that meant he probably owned, literally, the entire township, and part of another. Lemuel Milk may have owned around 20 percent of Newton County, and was one of the largest landowners in the state.

In April 1873, *The Kentland Gazette* reported that Indiana's attorney general was going to initiate a suit in Newton County court to recover around eight thousand acres of Beaver Lake land.[74] This was land bought by Milk, Cooley, and Dean from Michael Bright. Nothing seems to have come of this, however, and Milk and Cooley continued to develop their Beaver Lake holdings. In 1874, they extended Beaver Lake Ditch another mile into what was left of the lake, and held a party for the letting of the water into the ditch. Michael Bright and Algy Dean were in attendance, toasts were made, and R. D. Parsons—identified as "ditchest"—delivered a valedictory speech. The *Kentland Gazette* included the speech in a fawning article about Milk and Cooley, who were extolled by Parsons as the lake's "energetic and preserving captors . . . as their liberality has been so universally felt in our midst." Parsons addressed the lake in bombastic, mock-heroic terms:

> Beaver Lake farewell! The hand of progression and oppression has been laid heavily upon thee. Thy vast and heaving bosom has been bared to the scorching rays of the sun and the pitiless winds. For untold ages thou hast rolled thy heaving waters o'er this broad basin which Nature has made to hold them . . . For ages thy waters have been a resting place for aquatic birds in their semiannual migrations from the great north lakes to the sunny south and back. But now the time is soon to come when on weary pinion they shall seek thee and find thee not. . . . thou hast doubtless witnessed the rise and fall of mighty nations. . . . Alas! Poor lake, how changed thou art. On thy undulating bed waves vast fields of grain and grass, and o'er thy

serfdoms the dark green willow waves its pendant branches, and thy heaving, rolling waters, where are they? Today, reduced to the narrow limits of a ditch. Again, Beaver Lake farewell![75]

The article did not mention that the *Gazette* had only a year before reported that Indiana was preparing a lawsuit against Milk to recover land that actually belonged to the state. No person at that party probably ever heard of Ned Barker, who lived on the lake as a child, and reminisced in later years about the draining of Beaver Lake. The foolish pomposity of Parsons's speech should be answered by his remembrance:

> I saw the lake drain and millions of fish die as a result. Some survived in small standing pools only to be frozen out when winter came. Some of the buffalo fish weighed as high as 40 pounds. Geese and goslings could not fly in the summer and when they came out to drink, the eagles, mink and turtles preyed on the goslings. Some of the old birds tried to go on foot to other swamps. Some of us boys put big clothes baskets on horses and filled them with birds and took them home and watered them so they might live. We had to be careful to take the young along with the old to save all, or the old would return to the drained land and perish. Birds perished by the thousands and fish by the millions when the lake was drained.[76]

The *Kentland Gazette*'s ostentatious announcement of the new ditch to drain the lake suggests that Milk and his associates had little to fear that the state would actually sue them, or enjoin them from further development of Beaver Lake. They were probably correct. Indiana seemed unwilling to take Milk to court on its own, and the issue languished for five years. Finally in 1879, perhaps to avoid direct involvement by the attorney general, the state hired J. B. Julian as special counsel to handle the litigation against Milk and others, and Julian took the case to Newton County courts.

In 1881, Milk petitioned the legislature to have the suit against him dismissed.[77] Julian argued successfully against this petition, and though the proposal supporting it did not pass, another bill was introduced in 1889 to sell the Beaver Lake lands to current occupants for thirty-seven and a half cents per acre. By then the Indiana Supreme Court had held the Bright title invalid, which meant that the court had certified Indiana's ownership of the Beaver Lake lands held by Milk.[78] The legislature had to do something. Julian spoke strongly against the bill, arguing that the state was giving up land worth $80,000 for about $2,300, but in March the bill passed. Indiana divested itself finally of Beaver Lake, and a controversy that had begun as early as 1857 when Michael Bright claimed nearly sixteen thousand acres of lake bed, mostly underwater, was over.[79]

And what of the lake in 1889? The *Kentland Gazette*'s valediction for the lake in 1874 was premature; an 1876 map of Newton County shows a tiny sliver of water in the southeastern part of McClellan Township. An early history of the county (1883) says that the entire area of the lake except the southern part was under cultivation.[80] As late as 1889, newspapers still reported hunting parties bringing home game from "Beaver Lake," but by that year the lake was almost certainly gone.[81] Hunters may have found waterfowl on marshy areas that dotted the southern part of the township, but these small reminders of a once-large body of water would themselves disappear in a few years, as new techniques for draining, and better methods of financing it, became available.[82]

The story of the selling of Beaver Lake is a testament to the power of private property to usurp the interests of the public domain, even when the claims for ownership are bogus. Michael Bright used a specious title to claim Beaver Lake—a natural resource that belonged to all—in order to profit from selling land he never paid for. The state of Indiana enabled his scheme by legitimizing his claim, and used public money to begin the process of draining the lake. Bright and the state sold the lake to Lemuel Milk, Henry Cooley, and Algy Dean, who owned most of its land for twenty years, and completed the process of destroying the lake. Milk began selling his Beaver Lake lands in 1890; he died in 1893.[83] Michael Bright had died in 1881.[84] With their deaths, and with the final settlement of title in 1889, Beaver Lake as an issue, and as a potent source of drama that held public, legal, and legislative attention in Indiana for over thirty years, ended.

When Congressman Townsend warned in 1872 that land speculation endangered America's most beautiful lakes, he spoke for a growing number of Americans who saw a threat to lands whose value they believed transcended the pecuniary. Congress created Yellowstone National Park in that year, but Indiana was slow to follow suit. Over forty years later, in 1916, the state created its first state park at McCormick Creek in central Indiana.

Beaver Lake never had a chance to become a state park. Today about half of the old lake bed is owned by The Nature Conservancy, and is being restored as prairie. Efforts to "improve" Beaver Lake by draining it and transforming it into cropland were, at best, only partially successful. Farmers who bought land there often found themselves deeding it back to the original owner because the soil was too poor for profitable cultivation. In 1964, a farm manager for a Chicago bank visiting the area found that though some of the area was capable of growing corn and soybeans, "much of the land was very poor, [with] sandy soils and poorly drained. It was only good for pasture or 'to help hold the rest of the world together.'"[85] Michael Bright, Lemuel Milk, Henry Cooley, Algy Dean, and others no doubt made money by destroying the lake. But they did not improve its land. Today Beaver Lake Ditch is over twelve miles long and contains water most of the year. It still drains the Beaver Lake basin.

Beaver Lake and the marshes around the Kankakee River were home to humans for thousands of years.[86] Histories of the so-called "built environment" rarely provide an account of earlier inhabitants, and almost never take into account the biological entities that were extinguished in order for the built environment to exist. Every history of place takes for granted an absence, an obliteration, really, of a terrain, a natural area, an ecology that was itself a habitation. What is destroyed is seldom memorialized, even more rarely mourned. Standing on ground that once formed the shores of Beaver Lake, one surveys half-formed prairie, a few cultivated fields, strips of woodland, a distant highway, some areas whose invasive growth is neither prairie vegetation nor crops. Beaver Lake survives now only as a kind of symbolic presence, which must serve for this historical moment: In 2018, residents began a protest against a Texas dairy company that wanted to build a confined animal feeding operation (CAFO) on the lake bed. Signs sprouted up all over that part of the county which said "No CAFO on Beaver Lake." Something of the spirit of the old lake survived, and still called out for protection.

# NOTES

1. John A. Jakle, "Toward a Geographical History of Indiana: Landscape and Place in the Historical Imagination," *Indiana Magazine of History* 89, no. 3 (September 1993): 181.
2. Charles Test, "Beaver Lake, a Reminiscence," *Indianapolis Journal*, November 11, 1877. All newspaper articles, except as noted, were accessed online at Newspapers.com or Newspaperarchive.com.
3. Hugh Prince, *Wetlands of the American Midwest: A Historical Geography of Changing Attitudes* (University of Chicago Press, 1997), 137.
4. Prince, *Wetlands of the American Midwest*, 37.
5. S. C. Gwynne, *Empire of the Summer Moon: Quanah Parker and the Rise and Fall of the Comanches* (Scribner, 2011), 83.
6. Beth Bassett, "Land Surveying in Newton County, 1834–1835," *The Newcomer* 23, no. 1 (Winter 2017): 1–4.
7. From Adam Shidler's deposition in "The Report [to the Senate] of the Commissioner of the General Land Office Relative to the Drainage of Beaver Lake, Indiana, [1872]," 4. This document appears in *The Executive Documents for The Senate of the United States for the Second Session of the Forty-Second Congress, 1871–'72*.
8. Since the lake was not surveyed, all accounts of acreage were estimates. *The Madison Dollar Weekly Courier*, April 4, 1854, put it at fourteen thousand acres; Louis H. Hamilton

and William Darroch, in Hamilton and Darroch, eds., *The Standard History of Jasper and Newton Counties, Indiana* (Lewis Publishing, 1916), put it at "more than" sixteen thousand acres (224). The actual acreage probably varied with rainfall. Michael Bright was to claim the entire lake bed, which he platted at 15,760 acres in 1857.

9. Test, "A Reminiscence," *Indianapolis Journal*, November 11, 1877.
10. "Annual Report of the Auditor of State for the State of Indiana for the year 1854," in *Documents of the General Assembly of Indiana at the Thirty-Eighth Session* (Indianapolis, 1855), 235.
11. Stephen Strausberg, "Indiana and the Swamp Land Act: A Study in State Administration," *Indiana Magazine of History* 73, no. 3 (September 1977): 196.
12. "Annual Report of the Auditor of State of Indiana 1854," in *Documents of the General Assembly of Indiana at the Thirty-Eighth Session*, 236.
13. Leon M. Gordon II, "The Price of Isolation in Northern Indiana, 1830–1860," *Indiana Magazine of History* 46, no. 2 (June 1950): 152.
14. Qtd. in Strausberg, "Indiana and the Swamp Land Act," 197.
15. Strausberg, "Indiana and the Swamp Land Act," 196.
16. "Notice to Contractors," *Montecito Prairie Chieftan*, November 10, 1853.
17. "Notice to Contractors," *Jasper Banner*, February 16, 1854. Newspaper available at Newton County Historical Society, 310 E. Seymour, Kentland, IN, 47951.
18. Mary R. McCorvie and Christopher L. Lant, "Drainage District Formation and the Loss of Midwestern Wetlands, 1850–1930," *Agricultural History* 67, no. 4 (Autumn 1993): 28.
19. "Annual Report of the Auditor of State of Indiana 1854," in *Documents of the General Assembly of Indiana at the Thirty-Eighth Session*, 55.
20. Strausberg, "Indiana and the Swamp Land Act," 199.
21. "Report and Testimony of the Swamp Land Committee, 1863," in *Documents of the General Assembly of Indiana at the Forty-Second Regular Session, Part 2, vol. II* (Indianapolis, 1863), 1272.
22. *Report of the Swamp Land Committee to the General Assembly of the State of Indiana, 1859* (Indianapolis, 1859): 26–27. This report is available at the Indiana State Library, call number ISLO 336.12, no. 3.
23. *Report of the Swamp Land Committee*, 26–27.
24. *Journal of the Indiana State Senate, 40th Session* (Indianapolis, 1859), 1035.
25. *Documents of the General Assembly of Indiana at the Forty-Second Regular Session, Part 2, vol. II* (Indianapolis, 1863), 1272–96.
26. Strausberg, "Indiana and the Swamp Land Act," 202.
27. Shidler, "Report [to the Senate]," 4.
28. Shidler, in "Report [to the Senate]," 33.
29. *Report of the Swamp Land Committee*, 15–19, 20, 33.

30. From Amzie Condit's deposition in "Letter from the Acting Secretary of the Interior, in answer to a Resolution of the House of April 2, 1872, relative to Beaver Lake, in the County of Newton, State of Indiana," in *The Executive Documents of the House of Representatives, 42nd Congress, 2nd Session, 1871–1872*, Ex. Doc. 245: 6, available through Google Books, https://books.google.com.
31. "Amzie B. Condit," *Kentland Gazette* (Kentland, IN), April 10, 1873. Newspaper available at Newton County Historical Society, 310 E. Seymour, Kentland, IN, 47951.
32. [No Title], *New Albany Daily Ledger* (New Albany, IN), August 8, 1853, 3.
33. "The Report of the Commissioner of the General Land Office relative to the Drainage of Beaver Lake, Indiana," in *The Executive Documents of the Senate of the United States, 42nd Congress, 2nd Session, 1871–1872*. Ex. Doc. 25: 12. A complete record of Beaver Lake land sales from 1853 to 1871 is provided by the Recorder of Newton County in this document. Condit's remarks appear in his deposition in "Letter from the Acting Secretary of the Interior," 245: 6.
34. Among many others, the names of Bright, Dunn, and Condit appear on the list of ditching contractors in *Report of the Swamp Land Committee*, 28–30.
35. Reprinted in "Swamp Lands," *Delphi Dollar Journal* (Delphi, IN), April 4, 1854.
36. "Some Letters of Jesse D. Bright to William H. English: 1842–1863," *Indiana Magazine of History* 30, no. 3 (September 1, 1934): 370.
37. For instance, *The New Albany Daily Ledger*, April 12, 1854, reported that Bright had "bought 14,000 acres of swamp land in Jasper County . . . and designs draining the lake," and the editor of the *Madison Dollar Courier*, April 4, 1854, believed that Bright had purchased fourteen thousand acres of Beaver Lake land, and expected to realize ten thousand acres more from the draining of the lake.
38. "The Report [to the Senate]," 14.
39. Hamilton and Darroch, eds., *Standard History of Jasper and Newton Counties*, 224.
40. From a verbatim transcript of Michael Bright's recorded deed for Beaver Lake as it appeared in the Records Office of Newton County, from Beth Bassett, Newton County Historical Society.
41. Shidler in "Report [to the Senate]," 4, and "Report of the Swamp Land Committee to the General Assembly of the State of Indiana 1859," 20.
42. Hamilton and Darroch, eds., *Standard History of Jasper and Newton Counties*, 224.
43. J. W. Conner, "The Beaver Lake Lands," *Kentland Gazette*, January 4, 1872. Newspaper available at Newton County Historical Society, 310 E. Seymour, Kentland, IN, 47951.
44. "Death of M. G. Bright," *Madison Weekly Herald*, January 26, 1881.
45. "Report [to the Senate]," 14–15.
46. "Annual Report of the Auditor of State of Indiana to the Legislature 1860," in *Documents of the General Assembly of Indiana at the Forty-First Session, January 10, 1861* (Indianapolis, 1861), 239–44, http://www.archive.org/details/documentaryjourn1860indi.

All the information in this and the preceding paragraph about Bright's complicated land transfers are from this source.

47. The Auditor's Report of 1860, cited in the previous note, provides considerable detail about the circumstances for other Bright transactions involving land transfers but is strangely muddled about this one. The pretext for the conveyance of these lands to the state, and why the committee believed Bright's absurd title to be valid, remains an historical conundrum.
48. "The Report [to the Senate]," 52.
49. "The Report [to the Senate]," 4, 32.
50. From Isaac Myers's deposition in "Letter from the Acting Secretary of the Interior," 10.
51. Hamilton and Darroch, eds., *Standard History of Jasper and Newton Counties*, 202.
52. See depositions from Joseph S. Hartley and David S. Larkins in "Letter from the Acting Secretary of the Interior," 5.
53. "The Beaver Lake Trouble," *Cambridge City Tribune* (Cambridge City, IN), August 10, 1871.
54. See "A Rush for Reality," *Indianapolis Evening Journal*, September 29, 1871, and Hamilton and Darroch, eds., *Standard History of Jasper and Newton Counties*, 263.
55. Qtd. in "The Beaver Lake Trouble," *Cambridge City Tribune*, August 10, 1871.
56. "The Beaver Lake Trouble," *Cambridge City Tribune*, August 10, 1871.
57. Anonymous "X," "The Beaver Lake Lands," *The Indianapolis Journal*, November 14, 1871, 3.
58. Anonymous "Y," "Beaver Lake Lands Again," *The Indianapolis Journal*, November 29, 1871, 7.
59. Conner, "The Beaver Lake Lands," *Kentland Gazette*, January 4, 1872. Newspaper available at Newton County Historical Society, 310 E. Seymour, Kentland, IN, 47951.
60. Senator Daniel Pratt, Senate Journal, December 14, 1871. Senate Resolution, 42nd Cong. 2nd Session (1871).
61. The list shows that Bright spent $4,610 to buy the Beaver Lake shoreline; he sold his riparian claim for $20,047. This is a return on investment of over 400 percent.
62. "The Report [to the Senate]," 33–50.
63. "The Report [to the Senate]," 55–58.
64. Test, "The Beaver Lake Lands," *Indianapolis Journal*, February 26, 1872.
65. Algy Dean's deposition in the "Letter from the Acting Secretary of the Interior," 2.
66. For example, Senator Pratt, in his resolution asking for an investigation into the draining of the lake (S. Res. 42nd Cong. 2nd Session) referred to "improvements" made to the lake bed, such as fences, houses, etc.
67. "A Bill To Release To The State Of Indiana The Lands Known As The Bed Of Beaver Lake, in Newton County, in said State, S. 616, March 2, 1872," *Congressional Globe 1348*, 42nd Cong. 2nd Session (1872), *Heinline Online*.

68. Townsend, like many others, thought that Dunn had claimed half of the lake bed, with Bright claiming the other half. Records in "The Report [to the Senate]," 15, show, however, that only Bright made this claim.
69. The Yellowstone Park, House of Representatives Report. 26, US Congress, 42nd Cong. 2nd Session, 1872.
70. Washington Townsend, Chair, "Report on Beaver Lake, in Indiana, from the Committee on the Public Lands, March 19, 1872," in *Reports of Committees of the House of Representatives for the 42nd Congress, 2nd Session, 1871–1872*, report no. 37, 5–6.
71. Townsend, "Report on Beaver Lake," report no. 37, 5–6.
72. "A Bill To Release To The State Of Indiana The Lands Known As The Bed Of Beaver Lake, in Newton County, in said State, S. 616, 42nd Cong. 2nd Session (1873)," *Heinline Online*.
73. P. W. Gates, "Hoosier Cattle Kings," *Indiana Magazine of History* 44, no. 1 (March 1948): 24.
74. "Beaver Lake Lands," *The Kentland Gazette*, April 10, 1873. Newspaper available at Newton County Historical Society, 310 E. Seymour, Kentland, IN, 47951.
75. "The Draining of Beaver Lake—the Last Ditch," *Kentland Gazette*, September 18, 1874. Newspaper available at Newton County Historical Society, 310 E. Seymour, Kentland, IN, 47951.
76. Gerald Born, "Ned Barker Remembers Beaver Lake," *The Newcomer* 2, no. 3 (Autumn 1995).
77. Lemuel Milk, *The Petition of Lemuel Milk, Jane A. Milk and Henry H. Cooley to the General Assembly of the State of Indiana, in Reference to Quieting Title to Beaver Lake Land* (Indianapolis, 1881). This document is available from the Indiana State Library, call number SLI 333.1, no. I385B.
78. 106 Ind. 435 7 NE 379 State v. Portsmouth Sav Bank (May, 1886), *HeinOnline*, Dist. file.
79. State attorneys general may have wanted to distance themselves from this case, perhaps because of Milk's influence. J. B. Julian reports in "The Beaver Lake Lands, Supplementary Reasons Why Senate Bill No. 2 and House Bill No. 64 For the Relief of Claimants Should Not Pass," 1889, that some legislators believed that the attorney general took no interest in the case. Julian claimed this was a canard intended to mislead legislators about the true nature of the case. Julian's report is available from the Indiana State Library, call number ISLO 333.1, no. 9.
80. F. A. Battey, *Counties of Warren, Benton, Jasper and Newton, Indiana* (Chicago, 1883), 604.
81. "City Life," *Indianapolis Sun*, March 29, 1889, 4.
82. McCorvie and Lant, "Drainage District Formation and the Loss of Midwestern Wetlands," 29.

83. "Chain of Title on Bogus Island," unpublished document from Beth Bassett, Newton County Historical Society, 1.
84. "Death of M. G. Bright," *Madison Weekly Herald Herald* (Madison, IN), January 26, 1881.
85. "Chain of Title on Bogus Island," 2.
86. Sarah Surface-Evans, "Intra-Wetland Land Use in The Kankakee Marsh Region of Northwestern Indiana," *Midcontinental Journal of Archeology* 40, no. 2 (Summer 2015): 186.

3

# THE "GIGANTIC SWINDLE" OF 1869–1872

RIP the Kankakee Valley Draining Company

THE DESTRUCTION OF BEAVER LAKE, ALTHOUGH BEGUN BY THE STATE, was accomplished by private citizens who abided by the age-old principle that they had the right to dispose of their property in whatever way they saw fit. That they had acquired the property by very dubious means never really mattered, even after that fact had been established. The state effectively endorsed Michael Bright's phony title by accepting Beaver Lake land from him. By the time that was brought to the attention of the US government—which claimed the true title to the land—it was too late. The lake was almost gone, and Indiana never seriously disputed Lemuel Milk's right to complete its destruction. Underlying the state's tacit acceptance of the owners' destructive "property rights" was approval of the land's "improvement." Milk and his associates enriched themselves by draining the lake, and the government of Indiana was unlikely to question actions that allowed it to increase taxes on the Beaver Lake lands as those lands increased in value.

Beaver Lake was drained by open ditches—mostly by a single ditch, in fact—and draining was possible because the lake was forty feet or more above the Kankakee River. Years of hard labor were necessary to construct and maintain that ditch and others, but as Amzie Condit showed in 1855, the hydrology of the project was straightforward: connect the lake to the river and the lake would eventually drain. Draining the Great Marsh was an entirely different matter. The marsh was not only huge, comprising half a million acres, but the river at the heart of it was at nearly the same elevation as the marsh. The distance from the river's origin near South Bend to the state line was only about eighty miles, but the river's actual length, due to its many oxbows and meanders, was nearly 250 miles. The meanders, and the fact that the river and the marsh occupied a broad, sandy, nearly level plain, meant that the river had very little current. The marsh could not be drained by constructing ditches from the marsh to the river because in a real sense, the marsh and the river were the same entity.

The Indiana General Assembly was not eager to finance an enterprise as complex, and potentially expensive, as draining the Kankakee Marsh. In 1855, the legislature rejected a bill to straighten the river and drain the marsh.[1] The bill, sponsored by John C. Walker of LaPorte, not only set out a strategy for draining the marsh by straightening the Kankakee River, it also proposed a centralized authority for the project, a strategy that would be very attractive to future legislatures. As the *Indiana State Sentinel* explained:

> By [Walker's] bill, the Swamp Land Commissioners of the counties bordering on the river were made a board, and required to act in concert. His plan was to make the Kankakee the grand center of a complete system of swamp land draining. Experience has shown that this plan would have been not only more effectual, but far cheaper to the state, than the miserable plan adopted.[2]

"The miserable plan" referred to was the 1851 county plan, which provided little oversight of local swamp land commissioners, allowing widespread fraud throughout the state, as discussed in chapter 2. Undeterred by the General Assembly's rejection of his bill, Walker surveyed the Kankakee River at his own expense in 1859, and issued a lengthy report that was summarized by the *Indianapolis Daily Journal* in 1861. The article explained the relationship between the river and the swamp, and why draining the marsh into the river was not practical:

> It is all one swamp, and all one river. The river is only a little deeper place in the swamp, or the swamp is only an extension of the river. There are no banks, no limit to mark the separation of one from another.—Draining in such a place is manifestly impossible. A million of ditches in the swamp would only fill with Kankakee water up to its own level, and there the water would stay. The only chance of draining, therefore, lies in the possibility of lowering the water in the Kankakee so that the water of the swamps can run down into it and run off with the current.—This can only be done by creating a current in the river some way, for at present there is next to none.[3]

Walker's plan for creating that crucial current was to become the template for Indiana's efforts to drain the marsh for the next sixty years. Walker's survey had found that

> [the Kankakee River Valley] was very flat, and the bends [of the river] innumerable. Not exactly innumerable either, for they were counted, and found to be two thousand in the distance stated [200 miles]. This survey discovered the mode, and the only mode, of increasing the current, and lowering the water of the river so as to allow it to carry off the water from the swamps. That mode was to cut the channel straight, or as nearly straight as practicable from St. Joseph [County] to the State line. The removal of two thousand bends would remove two thousand dams that hold the water back.[4]

This plan was endorsed enthusiastically by the *Daily Journal*. Its article, framed as "A Chance for Speculation," estimated the cost of draining at $500,000 and calculated the profit to be made by those willing to buy land in the marsh and invest in draining it:

> If thoroughly drained the land at the average price of land in the country adjoining it would be worth $6,000,000. So says the estimate. The profit in the speculation would therefore be the difference between $6,000,000 and the sum of the cost of drainage and the present value of the land $1,400,000, or $4,000,000. This is certainly the biggest speculation we know of.[5]

The day after the *Daily Journal* article appeared, the *Indiana State Sentinel* ran an article praising Walker for forming a company to drain the marsh. The *Sentinel* was even more enthusiastic about the project than the *Journal*. According to the *Sentinel* article, the Kankakee River had "injured" the land by making it "valueless." The *Sentinel* did its own calculation of profit, and extolled the amount of money to be made from draining—"*four million six hundred thousand dollars*" (emphasis the author's). The article concluded with a strong call to action:

> We regard this enterprise as one of the most important to the people interested and to the State that has been set on foot for many years. Its consummation would bring into cultivation, in a few years, a district of country larger and richer in soil than Marion county. No stream in the United States, of similar length, has as much inexhaustibly rich bottom land. This soil is now practically worthless. It may be the most productive in the state. What is now a wilderness of water, mud, moss and grass, partitioned out among trappers and fishermen, may be made the garden-spot of northern Indiana.[6]

Disregarding the estimate of pecuniary value made earlier, the writer veered sharply into hyperbole, very likely in anticipation of possible objections landholders might have to paying for the project: "The object to be attained is so vast and incalculable in its results that the amount to be levied to accomplish it becomes insignificant."[7]

The company Walker proposed, the Kankakee Valley Draining Company, was to be incorporated under the "Drains and Levies Corporation Act" of 1852. After the legislature's 1855 rejection of his bill to drain the marsh, Walker saw that his plan could be implemented under this act, which provided for the creation of private companies to carry out the draining of wetlands. Walker was a visionary. Every effort to drain the marsh in the future would be done by private companies formed under state statute, and every project included dredging and straightening the Kankakee River. Walker was right, too, about lawmakers' reluctance to spend money on draining the marsh. The General Assembly of Indiana never allocated any funds whatever for a draining project on the marsh, in spite of the explicit provision of the 1850 Swamp Lands Act that money derived from the sale of wetlands be used for draining.

In 1861, however, some in the legislature were worried about Walker's plan to form a company, and wanted to repeal, or at least amend, the drainage act of 1852. George Frasier, of Wabash and Kosciusko Counties, defended the effort to repeal the law, declaring flatly that the company would "likely make war among the people there [in northern Indiana], if this scheme is allowed to go on." Frasier explained that the statute

> contains no provision by which the company can be compelled to cut one spade full of dirt . . . The amount of their assessments, under the act of 1852, becomes a lien on all the real and personal estate of every land holder whose lands are to be improved in value by their work, and that collection of this assessment would subject their whole estate to judgement and execution for the payment of these supposed improvements. And the result will be that this company will not only take to themselves all the land lying along the river, but perhaps otherwise injure the owners of this marshland.[8]

Bartlett Woods of Lake County agreed, pointing out that the law of 1852 "confers powers which no people should tolerate. It gives opportunity for the most gigantic swindles."[9] Others strongly supported the law of 1852, and resented Frasier's and Woods's suggestion that Walker's company could be a swindle. Marcus Packard of Marshall and Starke counties insisted that "the company was organized in good faith" and pointed out that "[l]arge tracts of their lands are overflowed by the river, which, by a little work, it has been discovered by competent engineers, can be made the most valuable lands in the country. I hope gentlemen will consider this. The best men in the county of Laporte are engaged in this, workmen who are above the imputation of any design of swindling."[10]

John Kenricks of St. Joseph County, who introduced the bill to repeal, argued for an amendment to the law, singling out precisely what landowners would most resent: arbitrary assessments. He believed that those should be made only with assent of landowners. In the end, arguments for amendments or repeal were not compelling. The bill to repeal the drainage act was referred to the Judiciary Committee, where it died.

Legislators' fears would have to wait ten years to become prophetic. Walker's company never materialized. The same month that its formation was announced (April 1861), he apparently changed his mind about the direction of his immediate future. Instead of pursuing the project to drain the marsh, he petitioned the governor to form a regiment of cavalry.[11] Efforts to drain the Great Marsh, and to find ways to pay for it, would have to wait until after the Civil War.

Wetlands may indeed have been as repugnant to settlers because of disease as many claimed, but this was not part of the discussions about draining the marsh in 1861, in the press or in the legislature. No mention was made either about the need for Indiana to create more cropland to feed its citizens. The *Daily Journal* makes explicit the most important reason for investing in the destruction of the marsh: "This is certainly the biggest speculation we know of."[12] Lawmakers would be eager to create legislation that increased that speculation, allowing the state to reap the benefits of higher taxation without having to pay for the improvements that enabled it.

The scope of work implied by Walker's plan was immense, though apparently no details were provided as to just how such an ambitious project would be carried out. In 1868, William C. Hannah, also of LaPorte, revived Walker's plan, and his company—or at least its name. Hannah's plan was detailed, comprehensive, and included an overhaul of the 1852 drainage law, which would get the full support of the General Assembly. Hannah's Kankakee Valley Draining Company advertised in its prospectus not only a clear purpose, but also a comprehensive scope of work. The company proposed

> [t]o reclaim or improve the lands of the valleys of the Kankakee River and its tributaries, by clearing out and deepening the present channel of said river in some places, and by excavating and opening new and more direct channels or drains therefor [*sic*] in other places, accordingly as the same may seem best adapted to economically and effectually drain said valley and reclaim and improve said lands.[13]

In addition to deepening and straightening the river, Hannah committed to constructing lateral ditches that would convey water from all "adjacent lakes, ponds, swamps and marshes" into the "improved" channel so that "injured" or "valueless" lands could be reclaimed. The company planned to alter the landscape of a piece of territory at least seventy miles long that lay in seven different counties:

> The proposed work is to be performed along the Kankakee Valley, between the place where the river crosses the line between the States of Indiana and Illinois, and a point on said river in St. Joseph county, to be hereafter determined by a survey, and will extend through or into the counties of Lake, Jasper, Porter, La Porte, Starke, and St. Joseph—the distance between the terminal points being about seventy miles in a straight line. All the lands believed to be liable to be affected by the proposed work lie within the territory embracing the counties aforesaid and Marshall county—portions thereof being situated in each of said counties.[14]

Hannah organized his first company under the 1852 law (amended in 1867), which gave drainage companies the right to hold liens against assessments, and to foreclose on property whose assessments were not paid. This was the same method that Walker's company would have used, which so disturbed some legislators in 1861. But Hannah wanted legislation that granted his company broader powers, especially for raising funds, and in 1869, he appealed to the legislature, which obliged with a much-expanded law.[15] The Drainage Act of 1869 was a comprehensive, 3,500-word document that focused on reclamation of the Kankakee River Valley (without actually naming it) and repealed all previous drainage legislation. With this repeal, the General Assembly made the 1869 law the centerpiece of its legislative efforts to drain the Kankakee Marsh, and all wetlands in Indiana. The act provided Hannah's company with the legal framework it needed

> for the purpose of draining, reclaiming and protecting such lands, which shall have power to straighten, widen, deepen, and make new channels for the whole or any part of any river, or water course, and to construct any dikes, drains, levees and breakwaters, and to do every thing which they shall deem proper, to accomplish the purposes for which the company shall have been organized.[16]

The law spelled out in detail how a drainage company should be formed, the method for choosing appraisers in each county, how assessments and reassessments should be made, and that assessments would become liens held by the company. It also authorized the company to appropriate any local resources it needed and to issue bonds secured by liens on property, and it allowed the payment of assessments in installments. The law was given emergency status, so it could take effect immediately. The Senate passed the bill unanimously,[17] the House by a large majority with little debate.[18] Pierce, of Porter County, probably spoke for many in the legislature when he argued that

> the bill has been carefully examined. It passed the Senate after having passed the scrutiny of the Judiciary Committee of that body, and I understand that it has the endorsement of the Judges of the Supreme Court. If the bill succeeds, it will assist in the development of a large amount of swamp land that is now valueless.

He added, "[w]e want the swamp lands reclaimed. Their reclamation will increase the taxes of the State."[19]

Only one representative, Jonathan Lamborn of White and Benton Counties, spoke strongly against the bill, echoing concerns that some lawmakers voiced in 1861. Lamborn maintained:

> this bill is a scheme for plunder; that its enforcement would rob perhaps hundreds of poor men of the State of their homesteads. It is well known that in the swamp land

districts there are many occupants of small farms who have secured places which they call their homes; and it is well known that many of these have small debts hanging over them; and, if this bill becomes a law and gives those companies power to assess these small farmers for their drainages, they will not be able to pay, and their homes will be swept away from them ... This bill was a scheme for capitalists to absorb the lands of the people; that all such legislation is wrong and repugnant to the instincts of a republican people, and that the operation of this bill will certainly work great hardship against the poorer class of the people of the State.

Lamborn spoke prophetically about the public reaction to the bill, saying that "the tendency of such legislation is to make enemies."[20] Events would very soon prove his prediction understated.

The bill became law in May 1869 and the company lost no time in beginning its work. In June, Hannah launched the first phase of the plan for draining the marsh: raising money. He published announcements in local papers that gave notice to property owners along the Kankakee River that his company would begin examining lands for assessments to pay for draining.[21] True to the schedule outlined in the company's prospectus, Hannah's strategy was to begin the work of choosing lands for assessments starting at the state line in Lake and Newton counties, and working northeast all the way to St. Joseph County. Examination of lands would begin on July 1 at the state line.[22]

Reaction to the commencement of the company's work in Newton County was swift and defiant. In August, several property owners, led by Algy Dean who owned land on Beaver Lake, met and voted to oppose what they called the "aggression" of the company. In a list of angry resolutions bordering on open revolt the group said it would

> [l]awfully resist, to the extent of our ability, any and all acts of said company to enter upon our lands with a view to ditch or drain the same without our consent, [and would] resist the payment of any assessment made by said company upon our lands, to the utmost legal extent of our ability, and if need be will use physical force to prevent the same.[23]

The citizens took issue not only with the company, but with the law that authorized it, claiming that it was "unconstitutional, and unprecedented in the annals of law-making, even under monarchical government."[24] The "war" foretold by Representative Frasier in 1861 had begun.

The charge of unconstitutionality was quickly addressed—possibly by the company itself. In September, James Bradley of LaPorte filed a complaint in Newton County

Circuit Court on behalf of his client, James O'Reiley, against the company.[25] The circumstances surrounding the case were suspicious. Bradley was a state senator from LaPorte and Stark counties, and had been instrumental in getting the 1869 drainage law passed.[26] James O'Reiley was not a resident of Newton County, and never appeared in court.[27] The *Newton County Democrat* called the case "fictitious," claiming it was manufactured by the company to test the constitutionality of the act.[28] *The Standard History of Jasper and Newton County* concurs, stating that many believed the suit was "collusive," designed to give the company the benefit of a favorable court decree.[29]

Several Newton County landowners who were part of the group sworn to oppose the company were permitted by the court to become parties to the suit. The plaintiffs asked for an injunction to stop assessments against their property on the basis that the law authorizing the company was unconstitutional. The judge in the case, Charles Test, refused to issue an injunction, and held that he had no authority to act in opposition to the legislature. However, he

> [a]dmitted that the law was *bad* [emphasis in the original]—that it was tyrannical—that it would act injuriously to the people. He "thanked God that he did not assist in making the law," and said: "If ever the people of this country lost their liberty, it would be through the legislative department of the government!"[30]

The plaintiffs appealed the ruling to the Indiana Supreme Court. In November 1869, the court reversed the Newton County Court's decision, and the lower court enjoined the company from further operations in Newton County.[31] However, the Supreme Court's lengthy opinion (which cited several drainage cases it had upheld) had something for everyone. The court did not declare the law unconstitutional, and upheld the legislature's basic right to create whatever drainage legislation it wanted.[32] As far as the Kankakee Valley Draining Company was concerned, the court's decision meant that it had been completely vindicated. In its 1871 prospectus, the company specifically cited the Newton County decision as having "fully sustained the constitutionality and validity of the law of 1869."[33]

Opposition to the company, and to the law, spread rapidly. A meeting similar to the one in Newton County was held the same month in St. Joseph County, and some local papers began denouncing the company as a "swindle."[34] By 1871, angry landowners in the northwestern part of the state wanted the law repealed, and petitions from Lake, LaPorte, Porter, and St. Joseph counties were presented to the legislature.[35] The petition from LaPorte summarized many citizen complaints, denouncing the law as "unjust and oppressive" because the property of citizens was placed at the mercy of a corporation, unrestricted as to geographical limit, and without authorized plan or mode of procedure for the condemnation of private rights affected. The act makes no provision for giving the owners of lands to be affected notice of the appointment of appraiser, or any voice in their selection, nor for security to the owner for the expenditure of the monies raised by assessment, nor for the maintenance of the levees, dykes and drains should they ever be constructed.[36]

The legislature declined to repeal the law. Instead, amendments were added, expanding its provisions. The Supplemental Act of 1871 required the company to file a map in each county showing all lands to be assessed, and required the posting of bonds to the state as surety for money collected.[37] Although believing that these additions and others would "cure defects in the [1869] law,"[38] legislators were determined to protect the company against the growing tide of public resentment. Section 8 of the act levied stiff penalties on anyone who tried to "obstruct or injure any construction" made by a corporation organized under the drainage law.[39]

The Supplemental Act became law in February 1871, and the company moved quickly to resume business. In April, it placed "a large amount" of bonds on the market[40] and in June, it accepted a bid from the J. J. Queally Company of Missouri for dredging the river. According to the report in the LaPorte *Argus*, the company had ten steam dredges that were ready to begin work.[41] However, the Supplemental Act did nothing to allay hostility to the company and to the 1869 law. Public outrage continued to grow. By August 1871, opposition meetings were held in Indianapolis, South Bend, LaPorte, and Valparaiso.[42] Newspapers that carried these stories wrote about the "gigantic fraud"[43] (*Richmond Weekly Palladium*) and the "great land-grab law"[44] (*Crown Point Register*), while a writer for the *South Bend Register* referred to officials of the Kankakee Valley Draining Company as "common thieves and robbers."[45] The story spread quickly beyond Indiana. The *Pittsburg Commercial* published an article alleging a "stupendous fraud,"[46] and the *New York Times* denounced the company as "a very gross swindle."[47] The *Chicago Tribune* condemned what it called the "infamy" of the Indiana legislature in legalizing a "gigantic land scheme" that would cost landowners over $4 million.[48]

The overheated rhetoric of these stories sometimes obscured the real grievances of landowners, who believed their property was appraised at inflated value and assessed at exorbitant rates. It should be emphasized here that the 1869 law allowed those assessments to become liens on the property—essentially mortgages owned by the Kankakee Valley Draining Company, which were used as security for the bonds it issued.[49] The *Crown Point Register* devoted five columns to the controversy in its story on August 31, reporting in detail the proceedings from meetings in Valparaiso and La Porte.[50] At the meeting in Valparaiso, citizens claimed that assessments were "so outrageous and absurdly unequal, unfair and unjust as to justify the belief that [the Kankakee Valley

Draining Company] is only a swindling concern, intending to defraud the people of their property under color of law." The "mass meeting" in LaPorte drew representatives from six different counties who condemned the company in a series of resolutions calling for, among other things, active resistance to the collection and enforcement of assessments and mortgages made on their lands. The representatives called for the immediate dissolution of the company.[51]

Even before these events, the company was beginning to buckle under the pressure of angry public meetings and the outcry in the press. Earlier that month, President Hannah resigned, and other members of the company withdrew from it, including James Bradley, the attorney for the company, who had strongly supported the 1869 law in the legislature and defended it before the Newton County Court in its first constitutional challenge.[52] The tide of bad press, and bad news, continued to rise. In September 1871 the *New Castle Courier* called the company's plan to ditch the Kankakee an "abortion,"[53] and in St. Joseph County, the company was declared in contempt of court for ignoring an injunction to stop assessments. The company's mortgages were voided, and attachments were ordered for former president William C. Hannah and vice-president J. G. Glidden.[54]

In November of 1871, George Cass, a founding director of the company, defended it at a meeting in Indianapolis. According to the *Indianapolis Journal*, about twenty men who owned land along the Kankakee River met with Cass and others to discuss the future of the company. A wealthy businessman, Cass was an important figure in various nineteenth-century commercial affairs, and in politics. Nephew of Lewis Cass (territorial governor of Michigan), he was a West Point graduate in engineering, the president of several different railroad companies, and had been a candidate for governor of Ohio.[55]

Cass spoke at length, explaining that work had been delayed because of court battles over the constitutionality of the 1869 law, even though the Indiana Supreme Court had upheld it, and by the attempt to repeal the law. He singled out what he called the "onslaught" against the company in the press that had brought negotiations for draining to a standstill. He claimed the company had been "very diligently at work," but admitted that no actual draining had taken place. He explained that another reason for the delay was the work of assessment, because so large an area had to be examined. He addressed that issue, about which there was so much bitterness, and conceded that the process might have been imperfect. He offered the following defense: "The objection against the inequality of assessment must go for nothing as long as men's judgements are fallible, except where there have been extravagant errors made."[56] That there had been many "extravagant errors," of course, was exactly the complaint made against the assessments.

Cass explicitly addressed a claim the company's critics had long held, which asserted that dredging the Kankakee River in Indiana would never allow a current strong enough to drain the marsh because of well-known impediments on the river in Illinois: a dam at Momence (six miles from the state line) and a rocky "ledge" nearby that slowed the current and was the cause of the impoundment of water in Indiana. (This claim is examined in detail in the following chapter.) Many believed that this ledge acted as a natural dam, and had actually caused the river to overflow in Indiana, creating the marsh. Illinois would never remove the dam, critics pointed out, and Indiana had no authority in Illinois that allowed it to destroy the ledge.[57] Cass refuted the idea that the natural dam was responsible for the marsh:

> The absurdity of the assertion that lands thirty and forty miles from the dam, and sixty to eighty feet above the top of the dam, could be overflowed by the action of the dam, should have convinced everyone that the statement was made to prejudice the public, and not for the purpose of giving reliable information.... The trouble of draining through the present channel is not on account of the fall, but the sinuosities of the stream, and the accumulation of drift wood and debris on account of the numerous bends.

However, the Kankakee Valley Draining Company recognized that the natural dam might be a problem, because there is "ignorance to contend with, as well as prejudice," Cass said. In anticipation of that, the company decided to become owners of the property bordering the ledge. More specifically, Cass pointed out, he had bought it: "I am the owner of two-thirds of that property, held for the benefit of the Kankakee Valley Draining Company, so that they can lower it, or remove it [the dams], or do with it whatever they please." The issue of the "rocky ledge" in Illinois was to become Indiana's bête noire for many years to come, well into the twentieth century, in fact, as discussed in later chapters.

Cass expressed confidence about the future of his company, claiming that "to some extent [it] has lived down the prejudice excited."[58] However, some at the meeting may have been uneasy when he added that the company would resume business as soon as it was "discharged from the courts," where it had been enjoined from conducting business. In other words, the work of the Kankakee Valley Draining Company had come to a complete halt because of ongoing litigation.

The following month (December 1871), William Hannah, former president of the company, defended it in the same newspaper, writing that the enterprise to drain the marsh

> [i]s not and never was, as asserted, a scheme to rob any farmer or other person, honest or dishonest, of any land or other thing whatever. It cannot be perceived what ground there is to base such an assertion upon, especially when a little investigation made in good faith into the affairs of the company (whose books and proceedings were at all times open to the public inspection, and all inquiries as promptly and fully

answered,) has never yet, it is believed, in a single instance, failed to satisfy the inquirer that the scheme is in fact intended only to benefit all the owners of wet lands in the Kankakee Valley, without injury, embarrassment, or injustice to any.

The issue of the natural obstruction in Illinois had been raised persistently, and Hannah addressed it, echoing Cass's rebuttal:

> It is not true, as asserted, that a natural obstruction in the Kankakee at Momence, or anywhere, prevents the successful drainage of the lands. If the concurrent testimony of several eminent engineers, who have examined it is reliable, the obstruction (if any) at Momence is not formidable, and whatever it may be, it is under the control of the company, and can be removed at small cost. It was always the intention of the company to remove it, if that should be found necessary to thorough drainage.[59]

By this point, there was little anyone could do to save the company. In 1872 county courts throughout the Kankakee Valley were overburdened with suits against the company—as many as four thousand, according to one claim.[60] In September 1872, the Porter County Court alone had 259 pending cases.[61] All this, in spite of the State Supreme Court's decision upholding the 1869 drainage law. By December 1872, pressure to repeal the law had again become intense in the legislature. In the House, Edwin T. Johnson of Marion County delivered a lengthy attack on the 1869 legislation, and the company;[62] his entire speech was reprinted word-for-word in the *Indianapolis Journal* the next day.[63]

The speech was a comprehensive, detailed critique of almost every section of the lengthy statute, and of the company's business practices. Johnson summarized the prevailing opinion of the legislation, claiming that the powers accorded by it were "unlimited in their extent, and monstrous in their character," allowing

> [a] few unprincipled speculators, residing perhaps in a distant State, happening to own a small tract of wet lands, or lands liable to be overflowed—lands which may be worth five dollars per acre, or which may be almost worthless, may, under the provisions of this law, take possession of the whole of a valuable water course, may absolutely control the great domain through which it passes, and may prostrate the interests of half a million of the people, and so long as the law remains no power can check them. Their power is bounded only by their own recklessness or avarice.[64]

Johnson's prolonged condemnation of the law, and the Kankakee Valley Draining Company's multiple abuses of it, did not include any criticisms of the legislators who had not only passed the 1869 law, but had expanded it in 1871, when they were petitioned to repeal it. Johnson also did not mention the decades-long support of legislators for the most hated provisions of the law, which allowed a private company to hold liens against the payment of assessments. He ended his nearly eight-thousand-word indictment of the company by declaring, "[w]e will break the back of this monster and let it sink in the swamps of the Kankakee." Repeal passed the House with only one dissenting vote;[65] the Senate concurred with an overwhelming majority.[66] The General Assembly's hypocrisy was acute: no one in either body acknowledged that the legislature itself had created the monster.

A year later, the board of directors met in LaPorte and effectively dissolved the company. In the colorful language of the *South Bend Tribune*, they took actions "which effectually [buried] that offensive carcass which [had] been a stench in the nostrils of our hard working farmers for the past three years."[67] The directors canceled all assessments and mortgages made by the company, and authorized the return of all bonds to the governor of Indiana. The company had not turned over a spade full of dirt. In its three years of operational existence, it did not accomplish any actual ditching of the marsh or dredging of the Kankakee River. Condemned in public meetings throughout the northern counties, derided in the press both in and out of the state, and vilified in the very legislature that had authorized its creation, the Kankakee Valley Draining Company was no more. The consequences of defeat for some of the company's supporters were severe. According to one local historian, former company president William C. Hannah was driven from his home in LaPorte, and representatives from northern Indiana who had voted for the 1869 law "were ruined for a time and could not have been more bitterly denounced had they been actual traitors to their country."[68]

The repeal of the 1869 drainage law and the ignominious demise of the Kankakee Valley Draining Company marked the end of the legislature's attempts to "corporatize" the project to drain the Kankakee Marsh. Resentment against the law and the company was based on what landowners saw as unfair practices by a corporation. Both the 1852 and 1869 laws authorized a private company to levy assessments—essentially taxes—against property, and to declare those assessments as liens owned by the company. When assessments were seen as capricious and unfair, landowners rebelled against legislation that they believed unconstitutionally privileged corporations.[69] The Indiana General Assembly would not make that mistake again.

The end of the Kankakee Valley Draining Company also marked the end of the ambition to drain the marsh using a centralized authority with sweeping powers to conduct work in multiple counties. This idea was one of the founding principles of the company, whose directors believed that a "general system" was required to drain an area as extensive as the Kankakee Marsh.[70] The "grand scheme" approach, with its promise of quick and inexpensive results, was no doubt attractive to legislators, but they lost little time in crafting new legislation that would require a much longer process for reclaiming the marsh but had the signal advantage of being far less controversial.

Three years after repeal of the 1869 law, the General Assembly passed a bill designed to bolster grassroot support for draining wetlands. The 1875 act empowered local landowners and county governments to initiate, finance, and control draining. Assessments were referred to as "taxes," and were limited to the expenses of draining.[71] This law was replaced and expanded in 1881 to provide a process for remonstrances (objections), not only to the locations of ditches, but also to assessments. Remonstrances were adjudicated by the board of commissioners, but could be appealed to the county circuit court.[72] This provision encouraged a drawn-out legal process carried out in multiple counties, which significantly lengthened the time required for draining projects to begin. This process will be discussed in the next chapter.

The drainage legislation of 1875 and 1881 sidestepped the issue of any obligation of the state to actually pay for the draining of the marsh, but in 1881 the government signaled its commitment to studying the problem by authorizing a comprehensive survey of the Kankakee Valley for the express purpose "of acquiring exact information and ascertaining and selecting the best, cheapest and most practicable outlets and routes whereby the wet and swamp lands of the Kankakee region" could be drained.[73] The man selected to conduct the survey was Professor John L. Campbell of Wabash College, Crawfordsville. Campbell's report was detailed, comprehensive, and meticulous. Campbell used the Beaver Lake area as evidence for the benefits of draining the marsh:

> The Beaver Lake region, in Newton County, is a good sample of the reclaimed marsh land along the Kankakee, and the results in this section are so satisfactory that the most earnest efforts should be made to recover all the overflowed lands.

Campbell echoed the confidence of John Walker, William Hannah, and numerous newspaper accounts about the huge bounty ready to be plucked from the reclaimed marsh:

> Owing to the favorable location of the Kankakee region with reference to the great commercial metropolis of the northwest [Chicago], and the facilities furnished by the numerous railways which pass through it, there will be a rapid increase in the value of the lands as soon as the drainage is effected. Estimating this increase at twenty dollars per acre, the aggregate addition to the wealth of the State will be eight million dollars ($8,000,000) on the estimated four hundred thousand acres reclaimed.

Campbell added that if the entire area drained by the Kankakee were to be reclaimed, the increased wealth to the state could be as much as ten million dollars. "Certainly," he wrote, "this is a problem worthy of the best efforts of the state."[74] Campbell's plan for addressing the "problem" was very similar to Walker's and Hannah's plans, but was more methodical. He specified three separate activities as steps in a process for transforming the marsh into cropland:

> First, the construction of a better main channel than now exists, for the flow of the river; second the straightening and deepening of the beds of the streams which empty into the main stream; and third, the digging of a large number of lateral ditches through the swamps to the improved channels.[75]

Fewer than twenty years later, this three-step method, carried out from county to county, would prove astonishingly effective.

Campbell seemed to be under the impression that the state would supervise the entire project, and divided the work into seven sections, beginning near the river's point of origin and ending near Momence, Illinois. He wrote, "[b]eyond the jurisdiction of Indiana it will be necessary to continue the improvement of the river to a point below the dam at Momence." He declared that the removal of the limestone ledge near Momence was a "necessity for the proper improvement of the river."[76]

Campbell explained that the ledge must be removed because the increased velocity of water in the straightened channel in Indiana will carry down "large quantities of soil and sand [into Illinois]." He pointed out that drainage into the Kankakee from Beaver Lake had caused the growth of weeds and grass in the river near Momence, and this would be increased to "a very damaging degree" unless an outlet is provided through the rock ledge.[77] Campbell didn't seem concerned about damage to the river downstream of the ledge's location if it were removed, but his warning here is the first clear, unequivocal admission that straightening the Kankakee River in Indiana would have serious environmental consequences for the Kankakee River in Illinois.

Despite Campbell's assumption that the state would supervise the vast project of draining the marsh, the General Assembly was in no mood to finance such an undertaking. In his address to the legislature in 1883, Governor Porter made a plea for funding the project that echoed Campbell's report and many newspaper articles, pointing out that "however considerable" that cost may appear, it "bears no sort of proportion to the additional value which drainage will impart to the lands." He admitted that Indiana had not lived up to the terms of the Swamp Land Act of 1850, which obligated it to use funds from the sale of swampland to pay for drainage: "It must be confessed that this engagement was imperfectly kept," he said. He acknowledged the legislature's reluctance to allocate funds for draining the marsh, but pleaded for "some kind of liberal legislation" since the state had not kept its promise under the terms of the Swamp Land Act.[78] The General Assembly was not receptive. The next governor, Isaac Gray, repeated the request for funds in 1885.[79] Again, the legislature was unmoved. No legislation was passed that year, or any other year, for the purpose of draining wetlands.

The legislature did not repeat the mistake of 1869 in trying to direct reclamation efforts by a central corporate authority established without citizen consent. Instead, local support for the project was courted. In January 1884, a group of prominent citizens from LaPorte, South Bend, and Valparaiso met in Chicago to raise money for draining the marsh.[80] In March, a notice appeared in newspapers throughout northern Indiana inviting "all persons interested in the subject of the drainage of the marshlands" of the Kankakee Valley to a meeting in South Bend.[81] At this organizational meeting, about sixty landowners pledged to "form a company and petition for the straightening of the Kankakee River."[82] That year, steam dredge boats began to excavate lateral ditches in the marsh.[83]

Thanks to Campbell's report, Indiana was more certain than ever that removal of the rock ledge on the river near Momence was key to draining the marsh. In 1888, Senator DeMotte, of Porter and Lake Counties, offered a resolution to form a commission of Indiana legislators to confer with the legislature of Illinois for securing legislation necessary to remove the rock ledge.[84] His proposal died in committee, but it didn't matter. In 1889, Indiana was presented with a gift that would make cooperating with the government of Illinois irrelevant.

# NOTES

1. *New Albany Daily Ledger*, February 12, 1855. All newspaper articles, except as noted, were accessed online at Newspapers.com or Newspaperarchive.com.
2. "The Kankakee Draining Company," *Indiana State Sentinel*, April 3, 1861.
3. "A Chance for Speculation," *Indianapolis Daily Journal*, April 2, 1861.
4. "A Chance for Speculation," *Indianapolis Daily Journal*, April 2, 1861.
5. "A Chance for Speculation," *Indianapolis Daily Journal*, April 2, 1861.
6. "The Kankakee Draining Company," *Indiana State Sentinel*, April 2, 1861.
7. "The Kankakee Draining Company," *Indiana State Sentinel*, April 2, 1861.
8. *Brevier Legislative Reports, Volume V, 1861* (Indianapolis, 1861), 38, https://purl.dlib.indiana.edu/iudl/law/brevier/VAA8558-05.
9. *Brevier Legislative Reports, Vol. V, 1861*, 38.
10. *Brevier Legislative Reports, Vol. V, 1861*, 38.
11. "Cavalry Regiment," *Indiana State Guard* (Indianapolis, IN), April 27, 1861.
12. "A Chance for Speculation," *Indianapolis Daily Journal*, April 2, 1861.
13. *Prospectus and Articles of Association of the Kankakee Valley Draining Co. of Indiana*

(New York, 1869), 6. Available at Newton County Clerk's Office, Kentland, Indiana, Drawer 14, Cause 212.
14. *Prospectus and Articles of Association*, 7.
15. *Brevier Legislative Reports, Volume XIII, 1872* (Indianapolis, 1872), 239, https://purl.dlib.indiana.edu/iudl/law/brevier/VAA8558-13.
16. *Laws of the State of Indiana, Forty-Sixth Regular Session of the General Assembly, 1869* (Indianapolis, 1869), 82–88.
17. *Brevier Legislative Reports, Volume X 1869* (Indianapolis, March 3, 1869), 584, https://purl.dlib.indiana.edu/iudl/law/brevier/VAA8558-10.
18. *Brevier Legislative Reports, Volume XI, 1869* (Indianapolis, May 8, 1869), 192, https://purl.dlib.indiana.edu/iudl/law/brevier/VAA8558-11.
19. *Brevier Legislative Reports, Volume XI, 1869*, 192.
20. *Brevier Legislative Reports, Volume XI, 1869*, 192.
21. "To The Public," *Kentland Gazette* (Kentland, IN), June 10, 1869. *Kentland Gazette* available from Newton County Historical Society, Kentland, IN; "To The Public," *Plymouth Weekly Republican* (Plymouth, IN), June 24, 1869.
22. "To The Public," *Kentland Gazette*, June 10, 1869.
23. "The Draining Business," *Kentland Gazette*, September 1869.
24. "The Draining Business," *Kentland Gazette*, September 1869.
25. "O'Reiley vs Kankakee Valley Draining Company," Newton County Indiana, Clerk's Office, Drawer 14, Cause 212.
26. *Brevier Legislative Reports Volume XI, 1869*, 99.
27. Louis H. Hamilton and William Darroch, eds., *The Standard History of Jasper and Newton Counties, Indiana* (Lewis Publishing, 1916), 219.
28. "James O'Reiley vs. The Kankakee Draining Company," *Newton County Democrat* (Newton County, IN), October 7, 1869. *Newton County Democrat* is available at the Newton County Historical Society, Kentland, IN.
29. Hamilton and Darroch, eds., *The Standard History of Jasper and Newton Counties*, 219.
30. "James O'Reiley vs. The Kankakee Draining Company," *Newton County Democrat*, October 7, 1869.
31. Hamilton and Darroch, eds., *The Standard History of Jasper and Newton Counties*, 220.
32. *Reports of Cases Argued and Determined in the Supreme Court of Judicature of the State of Indiana, vol. XXXII* (Indianapolis, 1883), 169, https://catalog.hathitrust.org/Record/010422294.
33. *Prospectus of the Kankakee Valley Draining Co. of Indiana* (LaPorte, IN, 1871), 9. Document available at LaPorte County Historical Museum, LaPorte, Indiana.
34. *Rochester Union Spy* (Rochester, IN), August 20, 1869.
35. *Brevier Legislative Reports, Volume XII, 1871* (Indianapolis, 1871), 56, 72, 225, https://purl.dlib.indiana.edu/iudl/law/brevier/VAA8558-12.

36. *Brevier Legislative Reports, Volume XII, 1871*, 225.
37. *Laws of the State Of Indiana, Forty-Seventh Regular Session, 1871* (Indianapolis, 1871), 11–15, https://babel.hathitrust.org/cgi/pt?id=uc1.aa0003060811&view=1up&seq=20&skin=2021.
38. *Brevier Legislative Reports, Volume XII, 1871*, 343.
39. *Laws of the State Of Indiana, Forty-Seventh Regular Session, 1871* (Indianapolis, 1871), 14.
40. Reprinted from the *LaPorte Herald* in *Fort Wayne Weekly Sentinel*, June 12, 1871.
41. Reprinted in the *Marshall County Republican* (Plymouth, IN), June 29, 1871.
42. "The Kankakee Drainage Swindle," *Indianapolis Journal*, August 16, 1871; "The K. V. D. Co. Excitement," *Crown Point Register*, August 31, 1871.
43. "Gigantic Fraud—Democratic Swindle," *Richmond Weekly Palladium*, August 14, 1871.
44. "The K. V. D Excitement," *Crown Point Register*, August 31, 1871.
45. Reprinted in "The Feeling in LaPorte," *Indianapolis Journal*, August 26, 1871.
46. "Swamp Land Scheme," *Pittsburg Commercial*, August 26, 1871.
47. *New York Times*, August 29, 1871.
48. "The Kankakee Swindle," *Chicago Tribune*, August 15, 1871.
49. *Laws of the State Of Indiana, Forty-Seventh Regular Session*, 11–15.
50. "The K. V. D. Co. Excitement," *Crown Point Register*, August 31, 1871.
51. "The K. V. D. Co. Excitement," *Crown Point Register*, August 31, 1871.
52. "The K. V. D. Co. Excitement," *Crown Point Register*, August 31, 1871.
53. *New Castle Courier* (New Castle, IN), September 8, 1871.
54. "Railroad Setback," *Evansville Daily Courier* (Evansville, IN), September 10, 1871.
55. "A Busy Life Ended," *Ft. Wayne Weekly Breeze* (Fort Wayne, IN), March 29, 1888.
56. "The Kankakee Drainage," *Indianapolis Journal*, November 30, 1871.
57. "A Big Swindle," *The Indianapolis Journal*, August 15, 1871.
58. "The Kankakee Drainage," *Indianapolis Journal*, November 30, 1871.
59. William C. Hannah, "Kankakee Drainage," *Indianapolis Journal*, December 6, 1871.
60. *Brevier Legislative Reports XIII, 1872* (Indianapolis, 1872), 240, https://purl.dlib.indiana.edu/iudl/law/brevier/VAA8558-13.
61. *Brevier Legislative Reports XIII, 1872*, 240.
62. *Brevier Legislative Reports XIII, 1872*, 231–41.
63. "The Drainage Law," *Indianapolis Journal*, December 12, 1872.
64. *Brevier Legislative Reports XIII, 1872*, 232.
65. *Brevier Legislative Reports XIII, 1872*, 241.
66. *Brevier Legislative Reports XIII, 1872*, 223.
67. "Dead and Buried—The Last of the Kankakee Valley Draining Company," *South Bend Tribune* (South Bend, IN), November 11, 1873.
68. Thomas H. Cannon, ed., *History of Lake and Calumet Region of Indiana, Embracing*

the Counties of Lake, Porter, and LaPorte, vol. 1 (Historians' Association Publishers, 1927), 188.
69. "O'Reiley vs. Kankakee Valley Draining Company," Newton County Indiana, Clerk's Office, Drawer 14, Cause 212.
70. "The Kankakee Drainage," *Indianapolis Journal*, November 30, 1871.
71. *Laws of the State of Indiana, Forty-Ninth Session, 1875* (Indianapolis, 1875), 97–104, https://babel.hathitrust.org/cgi/pt?id=uc1.a0001996172&view=1up&seq=3&skin=2021.
72. *Laws of the State of Indiana, Fifty-Second Regular Session, 1881* (Indianapolis, 1881), 397–404, https://babel.hathitrust.org/cgi/pt?id=uc1.a0001996206&view=1up&seq=410&skin=2021.
73. *Laws of the State of Indiana Fifty-Second Regular Session, 1881* (Indianapolis, 1881), 561. https://babel.hathitrust.org/cgi/pt?id=uc1.a0001996206&view=1up&seq=7&skin=2021.
74. John L. Campbell, *Report Upon the Improvement of the Kankakee River and the Drainage of the Marsh Lands of Indiana* (Indianapolis, 1883), 15.
75. Campbell, *Report Upon the Improvement of the Kankakee River*, 15.
76. Campbell, *Report Upon the Improvement of the Kankakee River*, 23.
77. Campbell, *Report Upon the Improvement of the Kankakee River*, 23.
78. *Brevier Legislative Reports Volume XXI, 1883* (Indianapolis, 1883), 23–24.
79. *Brevier Legislative Reports Volume XXII, 1885* (Indianapolis, 1885), n.p.
80. *Madison Herald* (Madison, IN), January 25, 1884.
81. Fay Folsom Nichols, *The Kankakee: Chronicle of an Indiana River and Its Fabled Marshes* (Theo Gaus' sons, 1965), 188.
82. "Draining the Kankakee," *South Bend Daily Tribune*, March 17, 1884.
83. Nichols, *The Kankakee*, 188.
84. *Brevier Legislative Reports Volume XXIV, 1888* (Indianapolis, 1888), 160, https://webapp1.dlib.indiana.edu/brevier/view?docId=VAA8558-24&brand=brevier.

# 4

# STARTING A CURRENT, 1889–1893

## Indiana vs. the Rock Ledge

IN 1888, GEORGE W. CASS, FORMER RAILROAD EXECUTIVE AND FORMER MEMber of the board of directors of the Kankakee Valley Draining Company, died at his home in New York City. He was a prominent man from a prominent family, and his passing was noted in newspapers all over the country. A graduate of West Point, he had significant achievements in engineering, and had managed several railroad companies. The *New York Times* wrote that Cass had made a good deal of money out of railroads and had a fine reputation as a businessman.[1] He had been a candidate for governor of Ohio. The *Hope Pioneer* of Hope, North Dakota, said that many in eastern North Dakota would regret his death, and reminded readers that Cass County in that state was named for him.[2] His uncle, General Lewis Cass, had been territorial governor of Michigan, secretary of war under Andrew Jackson, and had been a key figure in Jackson's policies of Indian removal.[3] Cass County in Indiana is named for him. No obituary mentioned George Cass's connection to the Kankakee Valley Draining Company, and none pointed out that when he died, he owned thousands of acres of Kankakee marshland in Lake County, Indiana, and Kankakee County, Illinois.

In addition to his interests in railroad management and engineering, George Cass had another signal interest: he believed in playing the long game. For over sixteen years he had held on to property along the Kankakee River near Momence, Illinois, that he had bought while he was a member of the board of directors of the Kankakee Valley Draining Company. This property bordered the section of riverbed known as the "rock ledge" and contained a mill dam—property that Cass had purchased for the express purpose of allowing the Kankakee Valley Draining Company to do "whatever it pleased" with the dam and the rock ledge.[4] In 1886, Cass sold some of this property to the Chicago and Eastern Illinois Railroad. As far as Cass was concerned, he had kept the rights to the dam, and to the channel where the rock ledge was located. Upon his death, that property passed to his heirs, his son, Charles Cass, his daughter, Mary, and her husband, William Shelby. In 1889, the Cass heirs did something unexpected. They deeded this property to the state of Indiana, and the state suddenly found itself the sole proprietor of something that for so long had seemed completely out of its reach: the part of the Kankakee River in Illinois that, as far as Indiana was concerned, contained the major impediment to draining the Kankakee Marsh. For many years, Indiana had looked upon the rock ledge in Illinois (more about this below) as a major point of blockage for the current of the Kankakee River in Indiana. Many felt that draining the marsh was impossible without removing this blockage, which, they believed, would increase the current in the river. This, they believed, would allow the marsh to be drained.

The 1889 law authorizing acceptance of deeds to the property was specific about why Indiana wanted the property, what was on it, and where it was located:

> Large sums of money have been expended, and are still expending, in the drainage of said territory [the Kankakee Marsh], which drainage can not, however, be successfully accomplished, owing to the natural obstruction caused by the limestone ledge near Momence, below the State line of Indiana, said ledge rising abruptly seven feet and one-half feet above the bed of the river, and extending, with diminishing altitude, for a mile and a half below; and ... the revenues of the State would be largely increased, and the value of the lands still owned by the State in said region greatly enhanced, by the removal of said obstruction.[5]

The deeds were to "the lands forming the bed of [the Kankakee River], extending three hundred feet or more in width from the west boundary line of the state to and including said limestone ledge at Momence."[6] The language of the act makes clear Indiana's commitment to the doctrine that successful draining of the marsh could be accomplished *only* by removing the limestone ledge, and that such removal itself would somehow greatly increase land values and state revenues.

The law also provided for the appointment by the governor of a commission to oversee the work of removal, and appropriated $40,000 for the project. The gift of the rock ledge contained a catch, however. The act made clear that granting of the title was conditioned on Indiana's carrying out removal of the ledge within five years of November 1, 1887. In addition, it was very specific about how much of the ledge should be removed: it must be lowered seven and a half feet at its highest point, with a channel of two hundred feet at the top and twenty-five feet at the bottom.[7]

Charles Cass and Mary and William Shelby had good reason to give Indiana access to the rock ledge. They had inherited large tracts of marshland near the Indiana–Illinois state line. Removal of the ledge meant that the value of their lands could increase enormously. The deal was a fine speculation: the Cass heirs stood to make a good profit on land that would be improved at Indiana's expense.

Indiana was eager to get to work. The governor quickly appointed a three-man commission to oversee the work, and they met with the auditor of the state and the attorney general in Indianapolis on March 23, only sixteen days after the act was passed. The legislation did not mince words about the purpose of the commission: its official name was the "Board of Commissioners for the Removal of the Limestone Ledge in the Kankakee River." The name implied a blithe confidence in the success of the commission's project. However, events of the next four years would show the hubris of the legislature's ambition. The "ledge" referred to was a four-mile reach of the river in which water was flowing over limestone bedrock that lay just below the surface.[8] "Removing" it may have been possible—at least some of it—but at what cost? The title of the commission, and the legislation that created it, also made no reference whatever to the fact that the rock ledge was in another state. The text of the act suggests a certain arrogance toward Illinois, as it implies that Indiana somehow had the right to alter the landscape in that state. Indeed, Indiana saw "property rights" as the trump card that would prevent Illinois from challenging its access to the ledge. It would soon learn, however, that another party's claim to that property would seriously interfere with the project to change the current of the Kankakee River. Legal battles played out in court and the legislature's own actions kept the project on the brink of collapse for nearly four years.

A point of clarification should be emphasized here. No actual "ledge" existed. The sandy riverbed of the Kankakee in Indiana was quite different from the riverbed west of the state line. As mentioned earlier, the "ledge" referred to was actually miles of limestone rock that formed the bed of the Kankakee in Illinois. Removing all of this rock was impossible. What was usually meant by "removal of the limestone ledge" was, in reality, only a reduction in height of this bed of limestone near Momence, a reduction that many believed would allow for a swifter current in the Indiana part of the Kankakee. Just how much of a reduction was necessary is explored later in this chapter. The phrase "removal of the limestone ledge" (or some variation) is used in this chapter and elsewhere in this book because that was the language used consistently in Indiana and Illinois to describe the project of reducing the height of the limestone bedrock near Momence. Unfortunately, some took this phrase literally, and believed that a mythical stone in Illinois impounded the waters of the Kankakee in Indiana, and that its destruction was the indispensable condition for draining the marsh. Many of these individuals were officials in the Indiana government, but others, including many in Illinois, used this phrase carelessly.

The first act of the board was the election of John Campbell as president.[9] Campbell was a natural choice for the board, and for its president. The state based its preoccupation with the rock ledge on Campbell's assertion in the 1883 *Report on the Improvement of the Kankakee River* that the removal of the ledge was a "necessity for the proper improvement of the river."[10] Isaac Dunn was elected secretary. James Kimball was the third member. The board needed an engineer to supervise the project, and they decided to invite William Whitten, engineer for the city of South Bend, to fill that position. The third act of the board was to secure *all* trust deeds to the right of way in the river channel. The board had discovered that the deeds conveyed to the state by the Cass heirs did not contain the *full* right of way to the channel where the ledge was located. The Chicago and Eastern Railroad owned a bridge over the river at that point, and balked at having to remove the bridge to allow for destroying the riverbed. The railroad refused repeated requests to convey the right of way, and on June 28, 1889, it sent a letter to the board formally denying any consent to changes in the Kankakee riverbed.[11]

The Cass heirs had probably conveyed the deeds to Indiana in good faith, not knowing of the railroad's claim. When George Cass sold part of the island to the railroad, he had expressly held on to the rights to the waterpower, which included two dams. During negotiations for the sale, however, Cass had made statements in a letter that the railroad claimed gave it the rights to preserve the dams.[12] This was the basis of the railroad's claims to the dams; it also claimed riparian rights to the channel where the ledge was located.

The railroad had a sizeable investment in the area near the rock ledge. It owned part of the island in the river near the ledge, which it used for the recreation of employees. Removing the disputed dams, one at each end of the island, would be a severe disruption to employee use of the river around the island. The railroad also had an ice business at that location, and said that allowing the "improvements" proposed by Indiana would cost the company as much as $30,000 a year.[13] The board, which was poised to advertise for bids for the work on the ledge, was stymied.

The commissioners appealed to William Shelby, one of the Cass heirs, to confer with the railroad. In the meeting of November 30, 1889, the board learned that Shelby had told the railroad that he would have the dams "thrown open" and keep them open until the railroad agreed to "satisfactory terms" regarding the removal of the rock ledge.[14] Shelby removed the dams early in 1890, but his crude threat didn't work. The railroad replaced the dams and sued, and Shelby, along with Charles Cass and other landowners, were temporarily enjoined by the Kankakee County Court from interfering in any way with the dams.[15] The Cass heirs claimed they had the right to remove the dams, and the case went to trial in the Kankakee County Court. A decision by the court was set for September 1890, but was delayed until April 1891. The court found in favor of the heirs, but the railroad appealed to the Illinois Second District Appellate Court, and the court sustained the injunction while it considered the case. A final decision would not be made until January 1892. That is where matters stood by the end of 1891.

Nearly three years after Indiana had authorized funds for the destruction of the ledge, no work had begun. Indiana's project to destroy the obstacle to the transformation of four hundred thousand acres of marsh into rich farmland was now in jeopardy, not because Illinois had objected to the project, but because a railroad had made a plausible claim to "property rights."

No doubt neither the Cass heirs or the commission saw the irony in this state of affairs. Railroads in Indiana, because they could convey produce quickly to Chicago, were key to increasing the value of the marshlands. And now a railroad had stopped Indiana from removing the impediment to converting the marsh into farmland.

The board had other problems than the railroad, however. By 1890, the commissioners knew that the $40,000 allocated was not enough to cover the cost of the work specified in the act of 1889. Even though the injunction against Shelby was still pending in the courts, in September 1890, the board reviewed three bids for the work. Of these bids, only one was less than the $40,000 authorized by the legislature. The appropriation was for a channel two hundred feet wide at the top, twenty-five feet wide at the bottom, and seven and a half feet deep. William Whitten, engineer for the board, estimated that 112,990 cubic yards of limestone would have to be removed to meet the specifications.[16] One of the bids for the work was for $101,691, another was for $881,132, and the lowest bid—which the board was obliged by law to accept—was for $34,800.[17] In their report to the governor for 1890, the board explained that the disparity between the lowest bid and the others was because Emil Sirois, the contractor, was the "agent" of Shelby, Cass, and other landowners, who would "protect him from actual loss, while his bid remains within the limits authorized by law."[18] In other words, the Cass heirs and others were willing to make up the difference in the real cost of the project to Sirois. Sirois was notified of the acceptance of his bid, and that he must post a bond of $50,000 to secure the contract. Because of delays caused by the court case, the deadline for posting the bond was extended to May 1891.

Even after accepting the lowest bid, the board knew that it needed more money, and it was worried that the specifications for the work might need to be modified. In its report to the governor for 1890, it requested an amendment to the act that would allow it more flexibility in deciding the dimensions for the work, and it asked for a further appropriation of $5,000.[19] The legislature granted the request for more flexibility, but stipulated that the cost of any changes to the specifications must not exceed the original $40,000 appropriated. The amendment refused the request for $5,000 and, inexplicably, required an additional bond of $5,000 from each member of the board.[20]

The amendment caused a severe blow to the commissioners. In its April 1891 meeting, the board read a letter from William Shelby, who wrote that he and Charles Cass agreed that the amendment left "matters in worse condition" than they were before. He concluded the letter by announcing that neither he nor Cass "were willing to carry out the contract with Sirois." President of the board John Campbell felt the loss of Shelby and Cass keenly. He wrote back:

> From the contents of this letter the Commissioners infer that both Mr. Cass and yourself withdraw from any further responsibility in connection with the removal

of the stone at Momence, Ill. It is hardly necessary for me to say that this action on your part will be the death blow to the enterprise. We will wait according to our agreement until the 15th day of May for Mr. Sirois to perfect his bond,—after which date we shall feel at liberty to take whatever action we may consider for the best interest of Indiana.[21]

At its May meeting, the board drafted a letter to the governor informing him that the deadline for the payment of Sirois's bond had passed, and that "two of the largest landowners in the region [Shelby and Cass] . . . with whom our chief correspondence has been had in reference to the work, now utterly refuse to become sureties on the bond of Mr. Sirois." The letter went on to say that the board had considered whether or not to advertise for more bids. However, it had concluded that unless other landowners "should voluntarily come forward and in good faith agree to supply the deficiency between the state appropriation and the probable cost of the work, that any further expenditure in the matter is not advisable."[22] The letter ended with the reminder that the resignations of two board members would take effect that day. With the resignations of John Campbell, president, and Isaac Dunn, secretary, the Board of Commissioners for the Removal of the Limestone Ledge was officially dissolved.

The public record of the board's activities—which is extensive—does not provide a reason for Campbell and Dunn's resignations.[23] However the defection of Shelby and Cass, the legislature's refusal to appropriate the modest request for $5,000, plus the insulting demand for an extra $5,000 bond from each board member surely had something to do with it. A lingering dissatisfaction with the General Assembly may also have played a part. In his 1892 report to the Indiana Engineering Society, William Whitten, engineer for the Board of Commissioners, made comments critical of the legislature that echoed Governor Porter's comments in 1883—that the legislature had not contributed its fair share of the funds necessary to drain the marsh. Whitten remarked that "it would seem to be the state's *duty* [emphasis his] to at least assist in this great undertaking."[24] In a later report, Whitten hinted at chafing under the parsimonious restrictions of the General Assembly's funding when he commented on the difficulty of "doing $140,000 worth of work with a $40,000 appropriation."[25]

The board was not revived until October of 1891. The governor appointed Franklin Landers, former governor, and John Brown, a wealthy banker from Lake County, to fill the vacancies left by Campbell and Dunn. At the board's November 12 meeting, Landers was elected president, and Kimball, who had not resigned, was elected secretary. William Whitten continued as chief engineer. The board proceeded to make a new inspection of the river, and determined at its next meeting that the channel must be three hundred feet wide at the surface, twenty-five feet wide at the bottom, and seven feet deep, all dimensions at the highest point of the rock. One of these dimensions, the top width, was larger than the act of 1889 specified. In its December meeting, even though the rights to the channel had still not been settled by the court, the board decided to advertise for bids on the new specifications.

By its February meeting in 1892, the board had the bad news. In January, the Appellate Court of Illinois had reversed the decision of the Kankakee County Court, agreeing with the Chicago and Eastern Illinois Railroad that it held all rights to the channel where the rock ledge was located. After nearly three years of litigation, the railroad had obtained the legal right to halt any work on the limestone ledge by the state of Indiana.[26] The board shelved all work on bids, and decided to negotiate with the railroad directly.

Throughout this three-year period, many in Illinois had become seriously concerned about Indiana's intentions to cut a channel through the riverbed in their state. When the Illinois attorney general was asked by the Chicago *Daily Tribune* in August of 1890 what authority Indiana had to remove rock in the river near Momence, he said that Indiana had no authority whatever. However, he admitted that Indiana might have an agreement with individuals who owned land along the Kankakee to do such work. But, he said, "[a]ny legal proceedings for the purpose of condemnation would have to begin under the Eminent Domain laws of this state and [be] carried on in the courts of Illinois." The attorney general added that he had no official notice from Indiana about the subject.[27] A month later, the Indiana attorney general dismissed any possibility of a legal challenge by Illinois: "Illinois has nothing on which to base a case. A state can go into another and acquire real estate as an individual . . . The property owners in that vicinity gave Indiana the title to the land and Illinois has nothing to do with it."[28] The attorney general was bluffing. He knew full well that Indiana's "title" had been successfully challenged in court, and that the project to remove the rock near Momence might have to be halted. He was right about one thing, however: Illinois did not seem inclined to get involved.

Engineer Cooley, of the Illinois Drainage Commission, was worried about the effects of the rock removal on the Illinois River:

> These marshes [in Indiana] are underlaid with from 10 to 12 feet of sand. In the event of that limestone dam being blasted away and a channel opened for all that water, great quantities of debris would be washed into the Illinois River. Necessarily this would choke the stream from Utica to its mouth, where it is very narrow, and the grade is less than thirty feet in 230 miles. Millions of cubic yards of sand would be washed into the stream and the channel would be filled to such an extent that the danger from floods would be greatly increased.

When asked what could be done to prevent such an event, Cooley said, "[t]he most we can do is complain to the state authorities. . . . There is little chance of the work proceeding before the next session of the General Assembly. Injunctions would restrain progress until then. Legislation can be had before any harm is done."[29]

Engineer Cooley was both an alarmist and a prophet. There was little chance that the "limestone dam" would be "blasted away," opening a channel for all that water—although Cooley's hyperbole was understandable given Indiana's unwavering belief that "removal of the rock dam" was somehow absolutely essential to the draining of the marsh. It was easy to take such rhetoric seriously when it was repeated often enough. Cooley's language was perhaps calculated to spur Illinois officials into some kind of action. Illinois, however, took no official steps to stop Indiana from proceeding with the project. No injunctions were filed, and the legislature did nothing. Like Indiana's endorsement of Michael Bright's claim to Beaver Lake, Illinois's refusal to act was a tacit approval of the rights of property owners to make destructive changes to public waterways. Illinois did not challenge Indiana's right to remove part of the Kankakee riverbed, which, if carried out according to plan, would dramatically alter the flow of the river in the state.

For most of 1892, the work of the Board of Commissioners for Removal of the Rock Ledge was at a standstill. After repeated meetings with the railroad, the commissioners were unable to reach an understanding that would allow them to proceed with their work. On October 2, John Brown, who had been negotiating with the railroad, reported that he had still made no progress.[30] For some reason, however, the board was still optimistic. In spite of the bad news from Brown, the commissioners agreed to advertise for another bid, and accepted one from David Sisk of Westville, Indiana.

The Board's optimism seems to have been justified. Its continued pressure finally paid off and late in October, the railroad gave in and made a deal to allow the state of Indiana to have its way with the rock ledge near Momence. The railroad didn't seem to want much. The board agreed to relocate a dock the railroad owned, and after removing one of the dams, to build a new one further down river. These changes would cost Indiana $2,500. In addition, Sisk wanted $57,120 for the work.[31] The board needed more money. In its nearly four years of existence, it had spent just over $5,000 of the original appropriation of $40,000. The board had $34,886. It needed $25,000 more. In their 1893 report to the governor, the commissioners made their appeal:

> In the opinion of the Board the contract submitted is the best that can possibly be made, and the bond offered is sufficient to guarantee the completion of the work, the sureties theron being worth more than a million dollars; hence it is respectfully urged that the present General Assembly be requested to appropriate said additional sum of $25,000 and thus enable the completion of this important work, which would add greatly to the taxable wealth of the State, and save the citizens large sums of money that are annually being expended in efforts to drain the vast area of country effected by the obstruction sought to be removed.

The board not only reminded the General Assembly of the principal reason for allowing the additional funds—the enrichment of the state—it also pointed out, rather modestly, its own role in achieving the success of a goal which for nearly four years, had been in doubt, and it made a promise:

> The present Board of Commissioners have given the matter of removing the rock much careful attention, and through untiring efforts have finally succeeded in having the objections and obstructions removed that heretofore prevented the commencement of the work. And if the required assistance, here suggested, should be granted by the General Assembly the work can be completed within the present year.[32]

That the commissioners' efforts had been "untiring" was an understatement. Their efforts, in fact, had been relentless. They had kept their focus on one goal: access to the rock ledge, in spite of not having the expected clear title; in spite of what seemed an interminable lawsuit; in spite of the defection of William Shelby, their main ally and bondsman; in spite of the resignation of their president and secretary; in spite of the legislature, which balked at a modest appropriation and demanded an increased bond from each of them; and in spite of a court decision that denied them legal access to the ledge. Through all of this, they had persevered.

Perhaps the Indiana General Assembly saw it that way, too. Or perhaps, like the commissioners, lawmakers knew they had to take what they could get. The legislature promptly authorized the $25,000 requested.[33] The Board of Commissioners for the Removal of the Limestone Ledge had not given up, and they had prevailed. At their meeting on January 10, 1893, the board submitted to the auditor of Indiana a signed contract with David Sisk of Westville, Indiana, for the destruction of the limestone ledge in the Kankakee River near Momence, Illinois.

What they were getting for the $57,120 that Sisk wanted was quite a bit less than the original act had called for. Sisk's bid was for 83 cents per cubic yard.[34] Whitten had estimated that the original specifications would require the removal of 112,990 cubic yards of rock. Removal of that much rock at 83 cents per cubic yard would cost $93,781. The board, and Indiana lawmakers, were settling for less. Sisk's bid for removal of 68,819 cubic yards was about 40 percent less than Whitten's estimate. The legislature had already shown that it was unwilling to pay for what that estimate really cost, and no wealthy landowner, including Shelby or Cass, stepped up with an offer to pay for more. The original specifications—a channel two hundred feet at the top, twenty-five feet at the bottom, with a depth of seven and a half feet—had been a requirement of the Cass heirs for the gift of the trust deeds. Those specifications—like the five-year time limit—were now moot, since according to the Illinois Second District Appellate Court, the Cass heirs did not actually have title to the channel where the rock ledge lay. The state of Indiana was going to blow up part of the limestone ledge in the Kankakee River not because of a gift from the heirs, but because of a deal with the Chicago and Eastern Illinois Railroad. And they were going to blow up a lot less than they had originally specified.

Work began the following summer. In July 1893, the *Chicago Tribune* reported that contractors had arrived in Momence and were preparing for beginning the removal of rock within a week. The article also reported that "word had reached [Momence]" that the superintendent of the Kankakee Waterworks and authorities from Kankakee Insane Hospital had gone to Springfield to get the attorney general to stop the work.[35] If any such entreaties were made, they had no effect.

A week later, the *Tribune* devoted a column and a half to the story, including a map of the river showing the location of the proposed cut. The story reported fifty men at work, and said that opposition to the plan had increased at Kankakee, a city a few miles downstream from Momence. Individuals there were arguing that cutting through the ledge would "release immense quantities of sand" that would fill up dams at cities downriver, affecting the waterpower created by the dams. Others worried about destructive flooding once a channel had been cut through the rock. Their concerns were premature, but well-founded. When Engineer Cooley of the Illinois Drainage Commission said a few years earlier that changes to the Kankakee River would bring sand into Illinois—lots of it—and cause flooding along the river, he was correct. This would happen years later, however, after the river was straightened, and the marsh drained, as discussed in later chapters.

The *Tribune* article quoted Engineer Whitten, supervisor of the work for Indiana, who said that the removal of the rock was not essential to the drainage of the marsh, but that straightening the river in Indiana would be disastrous for Momence. If Indiana was not permitted to make the cut in the rock, he said, "Momence will be compelled to make it to avoid being washed away." Whitten claimed that landowners in Indiana, who owned more than one hundred thousand acres of marshland, were preparing to straighten the river, and that "no one in Illinois can prevent them draining their lands through a natural water course."[36] Whitten knew that fears about the consequences of Indiana's work on the rock ledge were unfounded. The plans called for a relatively small reduction of the limestone outcrop, which alone would not cause destructive sand flows or flooding. But he did know about the consequences to Illinois if the Kankakee was straightened in Indiana. He knew there would be increased flooding, and he knew quantities of sand and soil would be washed down the river into Illinois.[37] His arrogant indifference, and that of the Indiana attorney general, about the consequences to Illinois of Indiana's plans for the marsh, reflect the general disregard of their neighboring state by Indiana landowners and legislators throughout the process of draining in later years.

At their meeting on Monday, July 24, the Momence Board of Aldermen discussed the matter and declined to interfere with the work.[38] In August, the blasting began. The rock was drilled with holes seven or eight feet apart, and a dozen blasts fired at once by electricity. Several teams of horses and two trains of flat cars carted away the broken rock, which was used to cover local roads.[39] The work continued until December.

Misunderstandings about "the rock" continued. In its article about the completion of the project, *The Indianapolis News* reported that "the work is practically finished now, and the rock has been removed from the Kankakee River." The article pointed out that "about 67,000 cubic yards of stone" were removed, and a keen reader might have wondered if there was *more* stone.[40] Indeed there was. In fact, only 66,347 cubic yards of rock had been removed.[41] This was less than the amount contracted for, and much less than Whitten's original estimate of 1889. The act of 1889 called for a reduction in depth of seven and a half feet; the rock had been lowered only two and three-tenths feet, perhaps even less.[42] Indiana did not blow up enough rock. It would have to come back for more.

How important, really, was removal of *any* of the limestone "rock dam"? The Board of Commissioners for the Removal of the Limestone Ledge believed their work, however limited, was successful. In their final report to the governor in 1895, they had this to say:

> The removal of the ledge and the effect on the surrounding country so far has met the expectations of its most sanguine friends. It was not presumed that the removal of the stone would drain the country entirely, but that it would furnish an outlet and make draining practicable.

The "outlet" referred to was the brand new channel in the rock that would presumably allow water to pass over the rocky barrier near Momence more easily. Indiana engineers believed that this water, drained from the marsh in Indiana, would carry a load of sand that would not build up on the rocky ledge, thus allowing "practicable" draining in Indiana. But where would the sand go after it passed Momence? It could only go farther downstream, where it would collect on the river bottom, and on any shallow place in the river. The report went on to note that the removal of stone had had some effect: "Its effect on the Kankakee River for fifteen or twenty miles above Momence has resulted in a marked reduction and a shrinkage of ground [of the marsh?]. It is also seen that it is a great advantage in time of flood, as the water passes off rapidly."

The "marked reduction of ground" probably referred to some reduction of the marsh. But this effect was only seen a few miles east of the state line, which meant that in the vast majority of the marsh, there had been no effect at all. The report continued with a recommendation for a strategy to drain the marsh, a strategy first suggested by

John Walker in 1861, refined by the Kankakee Valley Draining Company in 1869, and made more methodical by John Campbell in his 1882 survey:

> The Board is of the opinion that if the river was straightened, the obstruction therein removed and the channels improved it would make it practicable to cut side ditches to the main channel, it being sufficient to carry off the entire body of water, and thus the entire country could be reclaimed.[43]

Engineer Whitten included his own report to the board, which suggested more work be done in Illinois in order to obtain the best results in Indiana:

> But before the full effect of the work that has been done, or that may be done, in the way of lowering the rock ledge can be utilized to the benefit of the lands in Indiana it will be necessary to provide an improved channel or improved channels from the ledge to the state line, a distance of about seven miles.

Whitten goes on to explain that there are obstructions in this part of the river that must be removed:

> Until these drift deposits (or morraines) are removed to a sufficient depth, or the river channel changed so as to avoid them ... the benefit that is expected to ensue from the lowering of the river at Momence, by way of scouring out of the bed of the river to a lower level, can not be realized to the lands lying in this State [Indiana].[44]

Whitten was much more pessimistic about the results of the board's work, apparently, than the board itself. His view was that no matter how much of the limestone riverbed was "scoured out," there could be no benefit to Indiana until the river between Momence and the state line was dredged. Whitten had discovered these additional obstructions during his work on removal of the rock ledge, but his statement here is consistent with his belief that removal of rock near Momence would have little effect. In an address to the Indiana Engineering Society in 1892, he had said that removal of the rock ledge would make "so slight a reduction" ... that it "would affect the surface at the state line but little, and above the state line would soon practically disappear."[45]

As an issue, the necessity for removing the rock ledge stayed alive among Indiana engineers. In a letter to the Indiana Engineering Society in 1894, John Campbell defended his language in the 1882 report, upon which the Indiana legislature had based so much in its 1889 act for ledge removal:

I used in my report in 1882 this language: "The removal of the obstruction at Momence is essential to the proper improvement of the Kankakee River." I think the thing has been done which ought to have been done, and if the Kankakee receives that attention in the future, which we know it will, every one will say that the first and proper thing was to remove this obstruction.[46]

That the "obstruction" had actually been "removed" was hardly accurate. Campbell does not address this question, but future legislatures would have to. At the time, the official report to the governor was that, somehow, removal of the rock ledge, although having little effect above the state line, had made draining the Kankakee Marsh "practicable." The question of what that meant would have to be answered by Kankakee Valley landowners.

## NOTES

1. "Funeral of General Cass," *New York Times*, March 25, 1888. All newspaper articles, except as noted, were accessed online at Newspapers.com or Newspaperarchive.com.
2. "Dakota," *Hope Pioneer* (Hope, ND), April 6, 1888.
3. Thomas Campion, "Indian Removal and the Transformation of Northern Indiana," *Indiana Magazine of History* 107, no. 1 (March 2011): 45.
4. "The Kankakee Drainage," *Indianapolis Journal*, November 30, 1871.
5. *Laws of the State of Indiana, Passed and Published, at the Fifty-Sixth Regular Session of the General Assembly* (Indianapolis, 1889), 291.
6. *Laws of the State of Indiana, Fifty-Sixth Regular Session*, 291.
7. *Laws of the State of Indiana, Fifty-Sixth Regular Session*, 291.
8. J. Loreena Ivens et al., *The Kankakee River Yesterday and Today* (Illinois Department of Energy and Natural Resources, 1981), 9.
9. "Minutes of the Board of Commissioners for the Removal of the Limestone Ledge in the Kankakee River," March 23, 1889. This document is available from the Indiana State Archives in Indianapolis.
10. John L. Campbell, *Report Upon the Improvement of the Kankakee River and the Drainage of the Marsh Lands of Indiana* (Indianapolis, 1883), 15.
11. "Minutes of the Board of Commissioners for the Removal of the Limestone Ledge in the Kankakee River," July 19, 1889. This document is available from the Indiana State Archives in Indianapolis.
12. "Minutes of the Board of Commissioners for the Removal of the Limestone Ledge."

13. William Whitten, "Kankakee Drainage," *Twelfth Annual Report to the Indiana Society of Engineers, Held at Lafayette, Ind., Jan. 12, 13, and 14, 1892* (N.p., 1892), 68.
14. "Minutes of the Board of Commissioners for the Removal of the Limestone Ledge," November 30, 1889.
15. *Annual Report of the Board of Commissioners for the Removal of the Limestone Ledge in the Kankakee River, 1890* (Indianapolis, 1890), 6.
16. "Minutes of the Board of Commissioners for the Removal of the Limestone Ledge in the Kankakee River," September 4, 1890.
17. "Minutes of the Board of Commissioners for the Removal of the Limestone Ledge in the Kankakee River," September 4, 1890.
18. *Annual Report of the Board of Commissioners for the Removal of the Limestone Ledge, 1890*, 8.
19. *Annual Report of the Board, 1890*, 14–15.
20. *Laws of the State of Indiana, Published at the Fifty-Seventh Regular Session of the General Assembly, 1891* (Indianapolis, 1891), 198.
21. "Minutes of the Board of Commissioners for the Removal of the Rock Ledge," April 17, 1891.
22. "Minutes of the Board of Commissioners for the Removal of the Limestone Ledge in the Kankakee River," May 15, 1891.
23. Records consist of the board's minutes, 1889–1893, its annual reports to the governor, 1890–1893, and Whitten's and Campbell's reports to the Engineering Society of Indiana, 1890–1895.
24. Whitten, "Kankakee Drainage," *Twelfth Annual Report*, 56.
25. Whitten, "Report of the Committee on Drainage," *Proceedings of the Thirteenth Annual Meeting of the Indiana Engineering Society Held at South Bend, Indiana, January 23, 24, & 25, 1893* (South Bend, IN, 1893), 20.
26. "Minutes of the Board of Commissioners for the Removal of the Limestone Ledge," February 2, 1892.
27. "Momence Rock Removal," *Chicago Daily Tribune*, August 2, 1890, 3.
28. "Kankakee River Commission," *Chicago Daily Tribune*, September 9, 1890, 6.
29. "Momence Rock Removal," *Chicago Daily Tribune*, August 2, 1890, 3.
30. "Minutes of the Board of Commissioners to Remove the Rock Ledge," October 2, 1892.
31. *Annual Report of the Board of Commissioners to Remove the Rock Ledge in the Kankakee, 1893* (Indianapolis, 1893), 13.
32. *Annual Report of the Board of Commissioners, 1893*.
33. *Laws of the State of Indiana, Passed and Published, at the Regular Session of the 58th Regular Session of the General Assembly* (Indianapolis, 1893), 56.
34. *Annual Report of the Board of Commissioners to Remove the Rock Ledge in the Kankakee, 1893* (Indianapolis, 1893), 13.

35. "May Change the Kankakee Channel," *Chicago Tribune*, July 23, 1893.
36. "Cut Away the Dam," *Chicago Tribune*, July 29, 1893.
37. See Whitten, "Kankakee Drainage," *Twelfth Annual Report*, 84, and John Campbell's letter to the Society, February 25, 1892, p. 93 in the same report.
38. "Cut Away the Dam," *Chicago Tribune*, July 29, 1893.
39. *The Noble County Democrat* (Albion, IN), November 2, 1893.
40. "Momence Rock: It Has Been Removed from the Kankakee River—The Results," *The Indianapolis News*, December 5, 1893.
41. *Report of the Board of Commissioners for the Removal of the Limestone Ledge in the Kankakee River* (Indianapolis, 1895), 18.
42. Whitten, "Report from President Whitten," *Proceedings of the Fourteenth Annual Meeting of the Indiana Engineering Society held at Indianapolis, Indiana, January 2, 3 and 4, 1894* (South Bend, IN, 1895), 19. The issue of how much rock was actually removed is complicated. A 1931 report to Congress by the US Army Corps of Engineers (*Report on the Kankakee River, House Document 784*, 41) states that in order to lower the ledge to the dimensions specified by the commissioners, over 240,000 cubic yards of rock would have to be removed. This is over three and a half times the amount of rock reported by Whitten in 1894. The report resolves this conundrum by estimating that the average cut was only about one-fourth of the maximum depth reported by the Rock Ledge Commission. This means that the ledge may have been reduced by fewer than seven inches overall.
43. *Report of the Board of Commissioners for the Removal of the Limestone Ledge*, 15–16.
44. *Report of the Board of Commissioners for the Removal of the Limestone Ledge*, 17.
45. Whitten, "Kankakee Drainage," *Twelfth Annual Report*, 71.
46. John L. Campbell, "How to Use the Waters of the Kankakee," *Fourteenth Annual Meeting of the Indiana Engineering Society held at Indianapolis, Indiana, January 2, 3 and 4, 1894* (South Bend, IN, 1895), 41.

# 5

# THE END OF THE MARSH, 1897–1923

## Landowners Straighten the River

THE YEAR OF THE LIMESTONE LEDGE REMOVAL, 1893, MARKED THE FORtieth anniversary of legislative discussion about the draining of the Kankakee Marsh. The first record of such discussion was in 1853, when representative John Walker proposed a resolution asking the governor of Indiana to open a correspondence with the governor of Illinois about draining the marsh. His motion was adopted. At that same session, Representative Spencer offered a motion to sell the swamplands at 75 cents an acre and use the proceeds to drain them. His motion was ignored.[1] Walker's resolution may have passed, but if such a correspondence was actually begun, it produced no results. Pleas to the legislature for funding from Governors Porter and Gray in the 1880s were fruitless. Newspapers sporadically reported other efforts to get the state to support drainage. For instance, a similar resolution to Walker's was adopted in 1877, but again, with no results.[2]

Engineer William Whitten, who celebrated his fiftieth birthday in 1893, was a keen student of the long history of discussions about draining the marsh, discussions that had gone on for most of his life. In his 1892 report on marsh drainage to Indiana engineers (more about this report in chapter 7), he commented on the forty-year history of such discussions. He admitted that very little had been accomplished in all that time, even though some landowners had taken matters into their own hands:

> It is true that the owners of some of these lands, despairing of the State doing its duty in the matter, have expended large sums of money for ditches, in some cases with satisfactory results, but in most cases resulting only in disappointment and loss. The work heretofore done has been local in character and without any general plan. In exceptionally favorable localities the fact has been demonstrated that the marsh when drained makes valuable land, but in a great majority of instances the money expended in ditching has been practically lost, and in some cases the work done has been an actual damage either to the person paying for the work, or to others.[3]

Whitten had an idea about how to improve this situation, and his idea was very similar to the one that John Campbell had proposed in his 1883 report on draining the marsh.[4] Whitten, like Campbell, advocated for a "general plan" of draining, and also like Campbell, believed that the state should create such a plan, supervise it, and share in the cost.[5] Whitten noted that Indiana had achieved some of those goals in 1869, when it passed corporate legislation that created the Kankakee Valley Draining Company. He devoted a page and a half to summarizing the history of the company, why it failed, and how its failure resulted in the repeal of the legislation. Whitten might have noted that the company might not have failed had the legislature agreed to share in the cost of drainage.

Had Whitten been a bit more analytical about the forty-year discussion on draining the marsh, he would have commented on the fact that although the legislature had shown repeated reluctance to authorize state funding, it had created a great deal of legislation devoted to drainage, beginning with acts in 1851 and 1852. After the repeal of the lengthy 1869 act, lawmakers had passed a major drainage law in 1875, and amended it in 1881, 1885, and 1889. In 1893, an act authorizing drainage districts was passed. The 1889 law was amended in 1895, and again in 1903. All of this legislation was designed to encourage and enable local authorities to organize drainage projects.

The pattern of all this legislation, including the earlier acts of 1852 and 1869, seems clear: the General Assembly wanted the Kankakee Marsh drained, but it did not want to pay for it, and it did not want the government directly involved in supervising drainage. The legislation of 1852 and 1869 allowed private companies to make assessments and contract for work. When this corporate legislation failed spectacularly in 1872, new laws were carefully crafted that placed all responsibilities for raising money and letting contracts with local landowners and county officials. Objections (remonstrances) could be registered with county courts, which made the final decision about the construction of ditches. The general intent of this kind of legislation also seems clear: lawmakers wanted the increased revenue from higher taxes on improved lands, but did not want to pay for the improvements. It hardly mattered that the Swamp Land Act of 1850 required the state to drain wetlands with proceeds received from their sale. Indiana ignored the requirement, and the federal government showed no interest in enforcing it.

In 1895, the General Assembly made its position clear on draining the marsh. In that year Senator Crumpacker proposed a bill to create a Kankakee River Commission that would oversee the straightening of the Kankakee River. The committee on swamp lands and drainage reported against the bill. Its stated position was that if the Kankakee Marsh should be drained, the work and expense should be borne by the landowners who would benefit. Crumpacker's bill died in committee.[6] As far as the issue of funding the draining of the marsh was concerned, the General Assembly was consistent. In 1855, forty years before, John Walker, perhaps realizing that correspondence between the governors of Indiana and Illinois had gone nowhere, introduced a bill to authorize an appropriation for straightening the Kankakee River. His bill failed twice in the same session.[7]

Full-scale efforts to straighten the river were at a standstill by 1895. The legislature may have expended $65,000 to destroy the rock ledge, but that would be the last state expenditure for modifying the river or draining the marsh. Gauging the actual effect of the rock removal on the river above the state line is difficult. The Board of Commissioners for the Removal of the Limestone Ledge may have reported that removal had "met expectations,"[8] but the report provided no details about its actual effect on the Kankakee in Indiana. Reports from the press between 1894 and early 1896 are sketchy and mixed. In early 1894, the *Chicago Daily Tribune* said that removal "produced such beneficial results," that marsh land owners were "more than ever enthusiastic" about draining.[9] However, in January 1896, the *Goshen Weekly News* noted that there was more water on the Kankakee marshes after a recent rain than at any time since 1869, and many had concluded that removal of the ledge had very little effect on the river above the state line.[10]

The Board had noted that removal would not have much effect on the river in Indiana, but it had been vague about the actual effect on drainage, and had been vague about whether or not removal would actually increase the current in the Kankakee River, which was widely believed necessary for successful draining. If landowners read reports, they may not have learned much. Many expected that removal would at least cause a faster flow of water in the Kankakee in Indiana, and some believed that removal itself would cause the marsh to drain. Whether or not removal of the rock ledge actually affected the Kankakee River in Indiana, especially near its source some eighty miles from the state line, is probably moot. A more important question is whether or not removal had any real effect on landowners' willingness to devote time, energy, and money to straightening the river.

That question, also, is hard to answer. It is a fact, however, as Whitten pointed out in 1892, that landowners in many parts of the Kankakee Valley were investing in drainage projects. These projects were undertaken regardless of the state's supposed removal of the rock ledge, and consisted of local dredging of channels into the river, rather than actual straightening of the river. Some were successful. As early as 1874, *The LaPorte Herald* reported that "excellent crops have been raised for a year or two in Lake County on the Kankakee Marsh, where ten years ago it was a great quagmire."[11]

Dredge boats were in use by 1884, and landowners were willing to invest in state-of-the-art technology. In 1885, the *Hagerstown Exponent* reported that a "new style of steam dredge" was bound for the marsh near Kouts, in Porter County. This dredge could dig a canal through marshy land over twice as fast as any other dredge then in existence. Its cutting boom carried iron buckets on a chain that emptied into a discharge boom at the rate of fifty per minute. The machine could excavate nearly two thousand cubic yards a day, and its body was made of pontoon sections, so that it could be transported by railway. Such a machine was designed to float in the canal as it was excavated, and it towed a floating boarding house for the crew. A man with a scythe mowed the marsh grass in front of the dredge.[12]

By the 1890s, sixty to seventy miles of ditches had been constructed in Lake County, paid for by a general assessment on all lands benefited by the drainage. These projects allowed for the production of large quantities of grain and vegetables.[13] Charles Landis, traveling in Jasper County in 1894, said that near DeMotte,

> [d]redges are working out immense channels that carry off the water rapidly. I saw land last week that could have been bought by the mile for one dollar per acre ten years ago that is now worth $25 an acre. Then this land was covered with water. Now the corn on it is up to a horse's back.[14]

Draining of the marsh had been going on at least since the 1870s, perhaps before. It is impossible to know the extent of these projects. Some were paid for by a single landowner on private land. Some were carried out under state statute, like the Singleton Ditch in Lake County, without litigation, so were hardly noticed.[15] Determining just when some ditches were begun or when completed is difficult. It is tempting to speculate that draining the marsh piecemeal may have resulted in creating productive farmland, at the same time allowing for large parts of the marsh to survive. However, the pressure to find better ways to drain more of the marsh and transform ever larger areas into the hyper-productive farmland that everyone believed lay just beneath its watery surface must have been intense.

No doubt the nearness of Chicago provided an important incentive for more draining. When William Whitten said that the city made draining the marsh "almost a necessity," he probably had in mind the need of the city's rapidly growing population for more food—food that the Kankakee Valley could provide. (The larger, metaphorical implications of Whitten's statement are explored later.) To be sure, Chicago offered a huge market for Indiana grain and vegetables. But it was Chicago's railroads that changed the landscape of the marsh—railroads that not only allowed much more efficient transportation of produce from farm to market, but also allowed produce to be transported easily to areas far west of Chicago, and far east and south of northwest Indiana. Railroads also brought the dredge boats to the Kankakee.

In his study of Chicago's importance to the economic development in the Midwest and the western US, William Cronon points out that by the late nineteenth century, Chicago had become the "principle wholesale market for the entire midcontinent." The city was at the center of a huge web of railroads radiating west, east, and south. By 1893, in fact, one-twenty-fifth of the world's entire railway mileage terminated in Chicago, a system that served some thirty million people.[16] An 1897 map shows that at least nine of these railroads crossed the Kankakee Marsh. These railroads provided not only quick access to Chicago, but also access to points west of the city, and east and south to Cincinnati and Pittsburgh. Markets all over the country were waiting and eager. The moment was ripe for agricultural expansion in the Kankakee Valley, and there were those who were determined to make the most of it.

Beginning in 1896, landowners in the eastern end of the Kankakee Valley began to organize with the express purpose not just for draining the marsh here and there, but for straightening part of the river, and draining a large section of marshland. The method worked out by these landowners was to channelize a few miles of the river, deepen streams that emptied into it, and construct lateral ditches from the marsh into the straightened river. Eventually, Indiana landowners all along the Kankakee would use this method for draining the entire marsh.

One of the first of these landowner organizations, and arguably the most important, was the Kankakee Valley Improvement Company, incorporated in La Porte County in September 1896. In January 1897, Dixon W. Place, president, published an open letter in a local newspaper about the company. Place was a well-known businessman, farmer, and marsh landowner from South Bend. His letter listed the names of the directors of the company (all local businessmen and landowners from nearby towns), explained the purpose of the company, and cited the law under which it was formed. The letter makes no mention of the removal of the rock ledge four years before:

> The object of this corporation is the improvement of the Kankakee River by straightening and deepening the channel, thus affording proper drainage for the land adjacent. The law under which we are incorporated was enacted by our legislature in 1889 and amended in 1896 [actually, 1895].

Place listed some of the provisions of the law, which directed a legal process that laid out the duties of the company, required a petition signed by a majority of affected landowners, and specified the role of county commissioners in appointing assessors and in issuing bonds. Assessments were to be paid in installments annually, the same as taxes. Place referred to the wet season of 1896, and used it as an argument for draining:

> In 1896 the very wet summer taught us that we cannot rely on Providence for our drainage but must do it ourselves.... [T]his wet season teaches us [Kankakee lands'] worthlessness if they are not drained. It has resulted in the effort to drain them and we ask the hearty co-operation of all parties to assist in making the Kankakee valley a garden spot of Indiana.[17]

Place noted the location of the project: the dredging was to begin a few miles south of Mud Lake in LaPorte County and extend about ten miles down the Kankakee.[18] This description of the ditching was modest. The Place Ditch—as it was to be called— was the first ditch to straighten part of the river. The plan of the project was very close to Campbell's proposal for straightening this part of the Kankakee in his 1883 report. It included straightening the main channel of the river and constructing a network of lateral ditches through the marsh that emptied into the straightened channel. The entire project was to comprise some twenty-six linear miles of ditches. The *Walkerton Independent* called it the "largest drainage waterway ever constructed on the Kankakee Marsh."[19] For the author of the *Twentieth Century History of La Porte County*, it was "by far the greatest undertaking; not because it was of greater magnitude and cost than others to follow it, but because it was the entering wedge of a new enterprise."[20] The "new enterprise" became the project to straighten the Kankakee all the way to the state line, and to drain four hundred thousand acres of wetlands. The Place Ditch was not only the "entering wedge" of the project to drain the entire marsh. Within a few years, Dixon Place would join a group of men to form an organization whose goal was not only the draining of the marsh, but the transformation of the business of farming in the Kankakee Valley. More of this later.

Place's project met considerable opposition. In his brief summary of the law under which his company was incorporated, he did not mention the elaborate system of objections to ditching allowed by the law, which listed no fewer than ten different options for remonstrances. Landowners could object if they felt assessments were too high, or if their assessments were high compared to others. They could object if they thought assessments were too high compared to benefits expected, or if their lands were assessed as damaged. The list went on.[21] Any remonstrance, if approved, had to be litigated, which meant that final approval of ditching was up to county courts. The Place Ditch would be challenged in three different county jurisdictions before it was allowed to proceed.[22] Resistance began immediately.

Within two weeks after Place made his argument for drainage in the Walkerton paper, C. F. Rupel wrote a rebuttal, claiming that drainage caused problems for some farmland. He wrote, "I believe with a great many others that the more you drain the water from the low land marshes, the dryer the high land becomes." Recent "long, dry summers," he said, were caused by draining "so many lakes, ponds, and low places" that no water was left to "cause any moisture in the air," thereby causing uplands to stay dry. He hoped that his readers should, "instead of making the Kankakee valley a garden spot of Indiana, keep the garden spots we already have."[23]

The next issue of the *Independent* reported that four hundred farmers and landowners had met recently in nearby Starke County to protest against drainage of the marsh. Their complaints echoed those made years before, in 1869–72, against the Kankakee

Valley Draining Company. The farmers claimed that more drainage would be "detrimental to what had already been accomplished," and that their hay crop was worth more than any other crop produced after further drainage. They protested against a special assessment for drainage, pointing out that the state had an obligation under the Swamp Land Act of 1850 to pay for drainage out of proceeds from the sale of wetlands. One complaint was particularly reminiscent of 1869–72:

> The farmers are particularly bitter against the two companies who have undertaken this drainage. They claim that Chicago land holders are trying to compel them to pay an extra assessment, and they will fight the matter through the courts. The drainage people are equally determined and a bitter contest will be the result.[24]

Bitterness on both sides emerged early. Dixon Place was kept busy throughout the spring of 1897 defending the advantages of draining. His debates with some of his fellow citizens in the Walkerton newspaper began amicably enough, but they did not stay that way. He responded to C. F. Rupel's claim that drainage caused dry spells by pointing out that "scientific study" had shown that idea to be erroneous. Place's carefully reasoned article showed a depth of knowledge about rainfall and soil moisture that put Rupel's anecdotal argument to shame. He provided statistics about rainfall in the Midwest from 1817 to 1865 that showed as many dry years as wet ones. He compared rainfall in areas near lakes to areas away from lakes that showed (he claimed) that growing crops promoted rainfall in dry areas. He insisted that well-drained land was superior to waterways in producing moisture. "It is a fact," he wrote, "demonstrated beyond successful contradiction that 100 acres of well tiled land planted in corn and well tilled will evaporate out of the soil and exhale by the growing crop more moisture during the growing season than will be evaporated from 100 acres of a lake." He cited specific data from the state agricultural station to back up his claim. He enlarged on the benefits of tile drainage, offering lessons from experience as evidence that well-tiled high land retained moisture better than land not tiled, and could withstand drought better. The benefits of tile drainage to lowlands were equally obvious, he claimed, explaining how tiles removed "surplus" water from wet land and provided "a warm, moist, airy seed bed three feet deep in which crops will take deep root."[25]

Rupel did not respond. But a week after Place's article appeared, B. F. Rinehart published a letter to the editor that changed the tone of the discourse. Reinhart, on the basis of his long experience as a farmer, agreed with Rupel that draining lowlands resulted in drought for uplands. He also referred to unnamed "leading scientists" who held that drainage caused less rainfall. He offered no evidence for this. His article went further than ideas about rainfall, however. He wrote:

The question may well be asked, what is the motive of the men who are making themselves so conspicuous, recently, in this matter of drainage? Is it philanthropy, or what is it? Now I do not wish to be understood as opposing the draining of lands that have need of it, but I am decidedly opposed to the way this work is being crowded. It is fast becoming a source of absolute oppression to a great many people.[26]

Rinehart's insinuation that the motives of the men who promoted drainage might be less than "philanthropic" struck home. In his reply, Place betrayed the real reason for the division between supporters of drainage and those against it:

They [men who want drainage] are actuated by a desire to increase the value of all Kankakee Valley lands and thus accomplish a public good that will prove a great blessing to coming generations. If all men were constituted like this Fossilman Rinehart there never would be any advancement. He has always taken his stand against every ditch or scheme of any kind looking to public improvement and persistently stands like a post stuck in the mud.[27]

The conflict was a familiar one. Young progressives who had studied the newest scientific ideas advocated for a change that older conservatives, citing years of experience, rejected. Dixon Place, and the men of the Kankakee Improvement Company, stood for the advancement of the public good, while Rinehart and those who agreed with him were obsolete "fossil[s]," who stood in the way of progress, like "post[s] stuck in the mud." A modern reader might be tempted to side with the old fossils, however. Place's unquestioned optimism that draining the marsh would prove a "great blessing to coming generations" would not be borne out in the next century.

Others joined the debate, with letters to the editor on the subject of drainage appearing in almost every issue of the Walkerton paper throughout early 1897. Those against "this drainage business,"[28] however, hardly understood the entire scope of the project begun by the Kankakee River Improvement Company, or the financial reach of the organization taking shape around them that was backing the project. Perhaps B. F. Rinehart had an inkling when he unleashed this bit of invective about the work of the company:

In my opinion it is one of the most complete systems of legalized robbery that ever was perpetrated upon a community. I look upon it as a whirlpool, a vortex, that is capable of swallowing up all the surplus wealth that the affected community can produce; and if this corporation succeeds in fastening itself upon this people, there will be hundreds who are now standing by, or who look with favor upon this scheme, that will agree with me.[29]

Opposition to drainage, and particularly to the Place Ditch, was strong, and for a while, seemed successful. Its first case was defeated in the LaPorte Circuit Court.[30] In June 1900, the Kankakee River Improvement Company got its second day in court in Valparaiso, Porter County. It was a big case, drawing over twenty-nine lawyers, most of them representing the remonstrants. Firms from Porter and LaPorte Counties, and cities as distant as Fort Wayne and Lafayette were represented. A former judge of the Indiana Supreme Court was also on hand. A lot of money was at stake. According to the *Starke County Democrat*, sixty thousand acres were to be assessed, for a total of over $174,000. The project was to straighten twenty-three miles of river, and include several lateral ditches, for a total of over forty miles, at a cost estimated at over $144,000. Almost two million cubic yards of marsh soil were to be excavated.[31] (These specifications were an exaggeration. See below.) The trial lasted a week, and the verdict went against the company. The project to remove the rock ledge on the Kankakee in Illinois played a significant part in the decision, the judge ruling that there was "not sufficient outlet" for the proposed ditch. The company was ordered to pay $5,000 in court costs.[32]

Unfazed by this setback, and the penalty, the company moved the case to the circuit court in nearby Knox, Starke County. The case was set for a hearing in March 1901, and there were reports that many citizens from LaPorte County, where the ditch had been defeated in court, were on their way to disrupt the hearing. No belligerent mob showed up, however, and the case was continued to trial.[33] Finding the right judge was key: On April 4, the Place Ditch was legally established by order of Judge Beeman, who had long been a supporter of draining the marsh.[34] The company had petitioned for the ditch in 1897; litigation in three different county jurisdictions had delayed construction for four years. Finally, in May 1901, a contract was let, and work began that summer on the first excavation for the straightening of the Kankakee River.[35]

Dixon Place was probably confident that his project would survive court challenges. He had found powerful allies whose financial backing would ensure the success of the Place Ditch, and other ditches. A month after the ditch was rejected by the Porter County Court, Place and others were preparing a petition to extend the ditch by five or six miles.[36] The men who signed the petition with Place were part of a group who would become the most important figures in the transformation of the marsh on the upper part of the river. This group was forming a syndicate—a consortium of companies that included the Kankakee River Improvement Company—whose goal was not only to straighten the river and drain the marsh, but to bring agriculture on an industrial scale to the Kankakee Valley. These men included Charles Tuesburg, Charles Danielson (vice-president of the Kankakee River Improvement Company), W. F. Cook, and W. W. Chapman, who, with several others, were officers or directors of the LaCrosse Land Company, the Tuesburg Land Company, and the McWilliams Land Company. These companies, with their interlocking directorships, eventually owned over fifty thousand acres of marshland. Their combined interests controlled forty miles of riverfront. The men who owned these companies were not speculators. They acquired land in the Kankakee Valley as a long-term investment.[37] Capital had come to northern Indiana.

The Place Ditch was completed in August 1902, draining marshland in southern St. Joseph, northern LaPorte, and Starke counties. The project actually consisted of two ditches. One, begun a few miles south of Mud Lake, straightened the Kankakee River for about nine miles. Another began in the marsh and emptied into the straightened channel. The ditch and its branches totaled about twenty-six linear miles.[38] The work took just over a year. Steam dredge boats made short work of muddy earth, acres of timber, huge swaths of marsh grass, and anything else in the way.

Even before the Place Ditch was completed, plans were made to straighten more of the river. In February 1902, Charles Tuesburg and others formed the Kankakee Valley Reclamation Company, whose goal was to dredge the Kankakee from the end of the Place Ditch to the border of Porter County, a distance of about seventeen miles through La Porte and Starke Counties.[39] Tuesburg and W. F. Cook had been constructing lateral ditches throughout properties in the marsh for two or three years; eventually they were to complete over fifty-five miles, with a goal of 125 miles by 1904.[40] Their proposal to straighten the Kankakee all the way to the border of Porter County did not meet the stiff resistance that hindered the construction of the Place Ditch. Remonstrances were not significant, perhaps because the syndicate owned much of the riverfront where ditching would take place. Another reason may have been that potential objectors were discouraged by the Kankakee Valley Improvement Company's major victory of 1901, which allowed the Place Ditch to proceed. In December of 1902, the company signed a contract with two dredging companies, one of which had dug the Place Ditch. W. H. H. Coffin, a director of the Kankakee Reclamation Company and superintendent on the Place Ditch, was appointed to the same position on the new ditch. Work began the following month.[41]

The company was in a hurry. The directors did not choose the lowest bidders for the work. The *Starke County Republican* explained that not only the company, but the citizens of the Kankakee Valley wanted the ditch as soon as possible. According to directors, those citizens would lose money without it:

> [T]he directors want the drainage as soon as possible. Each year without this ditch means the loss of perhaps two or three hundred thousand dollars to the people along the river, and the fear that [other bidders] could not do this work in a reasonable time caused the directors to contract with men on the ground and ready to go to work at once, at an advanced price.... [One] of the directors said, "[they] are competent dredge men and they are pushers... They have the machinery and will go to work at once, so the people along the ditch will have the benefits of the dredging, many

of them even next spring.... As it is, two large dredges will be at work, one of them within fifteen days, and the work will be pushed to rapid completion. What we want, and want bad, is for the ditch to be dug, and dug as soon as possible. Time is money to the people along the Kankakee River, and they can well afford to pay a little more money to have the work done thoroughly and quickly."[42]

The people of La Porte and Starke counties may or may not have been in a hurry to have their lands drained, but Charles Tuesburg's syndicate had plans for the reclaimed land, and it was quick to implement these changes. By 1904, the syndicate owned approximately seventy-six square miles in La Porte County, which had been extensively ditched, both by dredges and the installation of drain tiles. As soon as a section of land was drained (thirty-six square miles), each half-section was set aside for a single farm. On each farm, the syndicate provided an eight-room house, barns, and corn cribs, all set on concrete or brick foundations. Each house was painted and supplied with water. Fifty farms had been established in this manner by 1904. Tuesburg and his family lived on one of these farms for a while, before he built a house in LaCrosse. Tenant farmers were hired, and contracted to grow whatever crop the syndicate wanted. In the first year, the tenant paid one-third of the farm earnings for rent, and afterward two-fifths of the income. The syndicate also built roads, bridges, schools, stores, and fences. Capital had indeed arrived in northern Indiana, and its goal, as extolled by the author of the county history, was to make La Porte County "one of the garden spots of the world."[43]

Investment in the Garden Spot paid handsomely. The LaCross Land Company declared a 40 percent dividend on its investment for 1905, and announced a further 40 percent by March 1906. The article reporting this news included figures for acres of crops raised, and concluded with the rhetorical question: "Does drainage pay?"[44]

The Kankakee Valley Reclamation Company completed its ditch in February 1906, at the border where Starke, La Porte, and Porter Counties meet.[45] The ditch shortened forty-five miles of river to seventeen and took three years to complete. Even before it was finished, some predicted a bright future for ditching. The author of the history of LaPorte County heralded a 1903 amendment to the drainage act of 1889 that limited remonstrances, and placed more authority for processing ditching applications with county commissioners and county courts.[46] The author counted this a major victory for the forces of drainage, commenting, "Legitimate improvement has now received an impetus which cannot be checked.... Ditch after ditch will be dug, and the hitherto worthless lands of the low prairies will be redeemed and made productive."[47]

Ditches were, indeed, dug. In 1903, while the Kankakee Valley Reclamation Company was dredging the river in LaPorte County, the Miller Ditch was being constructed at the headwaters of the Kankakee, in St. Joseph County, near South Bend. Completed in 1904, this ditch was about seven miles long, and ended at Mud Lake, at the border of St. Joseph and LaPorte Counties, a few miles north of the Place Ditch.[48] In 1904 the Kankakee Valley Improvement Company completed a short ditch of five and a half miles from Mud Lake (which the Miller ditch had drained) connecting the Miller Ditch with the Place Ditch.[49] By 1906, when the Kankakee Valley Reclamation Ditch was finished, steam dredges had straightened forty-six miles of the river from its headwaters to the eastern border of Porter County, about half the length of the Kankakee Valley in Indiana.[50] Within five years, half the river had become a ditch, and approximately two-thirds of the marsh, over 250,000 acres, had been converted to farmland.[51]

The river was now a drainage canal in St. Joseph, LaPorte, and Starke Counties. Channelized through nearly half its watershed, it was surrounded by a huge network of ancillary ditches and dredged creeks that drained water, silt, sand, and other debris into the now swiftly flowing current of the narrowed stream. Landowners in three counties had accomplished a colossal feat of reclamation, but this triumph had created very large problems. Excess water, silt, and sand were overflowing into areas downstream of the Porter County line and into the newly dredged channel. Overflow from the Yellow River, the Kankakee's major tributary, was also a problem. Thirteen miles of the Yellow River in Starke County had been dredged by 1902.[52] However, near its mouth at the Kankakee, the dredged river bottom was much narrower than a few miles upriver due to erosion of the sandy soil in that part of the river. The narrow channel could not accommodate the sand and silt washing downriver, so was overflowing near its junction with the Kankakee. Sand deposits up to three feet were spreading across lands in that area.[53]

These problems were caused by the fact that projects to straighten the Kankakee in the upper valley were not only uncoordinated, they were quite haphazard. In the words of one report:

> The works of improvement have for the most part been done by local interests in a haphazard manner without due consideration of the problems involved or the effect of such improvements on other sections of the valley.[54]

The Kankakee River Reclamation Company began its ditch where the Place Ditch ended, but the Miller Ditch began miles upriver of the Place Ditch and was completed after the Place Ditch was dug. The Kankakee River Improvement Company did not complete a ditch connecting the two until 1904, two years before the Kankakee Reclamation Ditch was finished. The Yellow River was dredged in Starke County in 1902, years before the Kankakee Valley Reclamation Ditch was completed, which means sediment must have been collecting at the mouth of the Yellow for some time before the Kankakee was straightened at that point.

Such disorganized efforts had consequences. In their 1909 study of Kankakee reclamation, Engineers Downey and Pence commented on the problem of flooding in the lower valley: "The drainage operations in the upper valley have brought about a situation of the utmost gravity to owners of lands in the lower portion, where the river channel remains in its natural condition—shallow, crooked, and sluggish of flow."[55]

The "situation of utmost gravity" referred to deposits of sand, silt, and excess water flooding lands in Porter and Jasper, and perhaps even Lake and Newton Counties. To alleviate this, the report recommended straightening the river, but to be "essential to the full measure of success," dredging must proceed from the Porter County line all the way into Illinois. "Success" evidently meant relieving Indiana counties in the lower valley of flooding, but what, exactly, the "full measure of success" meant is not defined. They warned of consequences to Illinois if the river was only straightened in Indiana:

> If, however, the improvement of the Indiana division alone is constructed, the new channel will discharge into the river at the state line at the bottom of the present channel in the same manner as does the new channel of the upper section at the Porter County line.

This warning echoed William Whitten's threat in 1893: that if Indiana was hindered from removing the rock ledge, Momence would be "washed away" when landowners began straightening the river in Indiana. Downey and Spence were not optimistic about the possibility of a cooperative dredging operation between Illinois and Indiana. Whatever the consequences to Illinois, however, they fully recommended that dredging be completed in Indiana: "It is not probable that an interstate agreement can be reached for the construction of the Illinois division. In any event, the part of the channel in Indiana should and may be constructed as one project."[56]

By 1913, the consequences of dredging the river in the upper valley had grown worse, especially along the dredged part of the Yellow River, now known as the Elsbree Ditch. A report from the US Board of Engineers for Rivers and Harbors quotes a comment from an engineer who had inspected the territory in the Yellow River vicinity:

> The country is all cut up with ditches, and I make the assertion that no one of them is effective. The artificial channel made for the Yellow River has become an effective dam against it and is forced to find its outlet in the short reaches of its old channel and in the ditches on both sides of the Elsbree. The country is in effect a delta of the Yellow River and its many-mouthed outlet is through channels that were not designed and were never intended to carry it.... As the waters recede the drainage ditches are forced to take care of the Yellow River and perforce are unable to perform their proper functions. Yellow River is in effect "Lost River" and the entire system of drainage in the country is not worth maintenance.[57]

The entire length of the dredged Kankakee also had problems. Sand had accumulated on the bottom of the ditch—lots of sand. Originally about eight feet deep, the channel had been reduced to an average depth of four or five feet.[58] The upper Kankakee had to be re-dredged. In 1915, Dixon Place filed a petition to reconstruct the channel from the headwaters to the west line of Starke County. This thirty-nine-mile project, called the Dixon W. Place Ditch, would include widening and deepening twenty-six lateral ditches, and widening the Yellow River where it entered the Kankakee.[59] The project was litigated in the courts for years before it was authorized, and was still ongoing as late as 1923.[60]

As early as 1907, Horace Marble proposed a project that would implement Downey and Pence's recommendation to relieve flooding in the lower half of the valley. Marble was a recent convert to the idea of draining the marsh. A landowner and banker from Jasper County, he had grown wealthy from the sale of marsh grass, used for packing. Draining the marshes threatened his lucrative business. An ardent opponent of drainage, he criticized it during a gathering of drainage supporters in 1906. Charles Tuesburg stood up in the meeting and challenged him. According to the *Indianapolis News*, Tuesburg said,

> You don't have to drain if you don't want to. We have drained our lands up above you [in LaPorte and Starke Counties] and have found the result most satisfactory. We're tickled to death. If you fellows down here [in Porter County] want us to shove all of our water down here on you, why let it go at that. Go on raising wild hay, if you want to. We're raising corn up our way.

Marble was outvoted. The men at the meeting strongly supported draining, and Marble knew when to give up. He announced then and there that he would not "oppose the ditch any longer."[61]

Tuesberg's smug declaration of his satisfaction with the results of draining and his warning about flooding downriver exposed the ideological contradiction at the heart of a draining project that privileged private ownership above the public interest. Landowners upriver assumed that their ownership of riverfront property conferred the right not only to dredge the river, but to change its course—a decision that meant "shoving their water" on those downstream. Such a choice is essentially undemocratic, because it is made without the consent of *all* who are affected. Landowners downriver are presented with a fait accompli: they can either continue to be flooded, or dredge their part of the river, and flood others downstream of them. The contradiction lies in the double standard of those upriver, who claim the right to change river flow, but deny that those downriver have any rights in the decision. This contradiction, and the refusal of Illinois residents to dredge their part of the

Kankakee, was to bedevil relations between Indiana and Illinois for a century after draining in Indiana was completed.

Marble knew when he was beaten. He not only capitulated to draining, he took the lead in beginning the process for authorizing a ditch to drain the lower part of the Kankakee Valley. He filed a petition the following year in the Jasper County Court to straighten the river from the west line of Starke County through Porter County into Lake and Newton Counties. Called the Marble Ditch, this project would extend the channelized Kankakee almost to the state line. Protracted litigation halted construction until 1914, when the Indiana Supreme Court sustained a lower court's decision that allowed it to go forward. Construction of the Marble Ditch and the reconstruction of the Dixon Place Ditch meant that major dredging projects were happening simultaneously in the lower and upper Kankakee Valley. Between 1915 and 1923, dredge boats operated almost continuously on the entire length of the Kankakee River in Indiana, from near South Bend to the state line.

Horace Marble did not live to see his ditch authorized. By the time the Indiana Supreme Court upheld the Jasper County's decision authorizing the ditch, it had become the Marble–Powers Ditch. W. F. Powers had submitted the original petition for the ditch along with Marble. Marble might have found it frustratingly ironic that his ditch was strongly opposed for over seven years, mainly by John Brown and other landowners who owned thousands of acres in Lake County on which they pastured cattle. Marble's business had depended on marsh grass, and so did theirs. Draining the marsh meant the destruction of the plant that sustained their herds. These landowners fought hard to keep the marsh in its original state. According to the *Starke County Democrat*, the record of the case contained 1,358 pages—not including testimony of witnesses.[62]

The Marble–Powers Ditch would be twenty-nine miles long, and shorten seventy-seven miles of sinuous oxbow curves into twenty-seven miles of straight ditch.[63] The ditch and its many laterals would drain marshland in Porter, Jasper, Lake, and Newton Counties. Advertisements for bids that appeared in the summer of 1914 specified that work would start by December of that year and be completed by December 1, 1916. (It was not actually completed until 1921.) The McWilliams dredging company of Chicago was awarded the contract. The company decided that such a mammoth project was worthy of its newest, biggest machines.

By fall, the company had transformed the tiny village of Burrow's Camp, on the river in Porter County, into an industrial outpost, ready to assemble giant dredge boats for service on the river. The boats were transported in sections by rail to a depot near the river, then hauled overland. The *Starke County Democrat* described them as two "mammoth dredges, three times as large as the average dredge." The work involved the use of "two large timbers, 50 feet long and 36 inches across" brought from the state of Washington. Six weeks were required to assemble the dredges.[64] The town was booming. It acquired a new blacksmith shop, a store, garages, and several new houses. Len Burrows, mayor and namesake of the town, wanted to cash in on the new prosperity. He purchased a hundred acres nearby and laid them out in lots. In order to keep up with the "progress of the town," and perhaps, to certify his new-found status as real estate magnate, he bought a car, "a swell new Maxwell."[65]

In spite of the huge dredge boats, construction of the ditch was delayed. Its progress was already a year behind schedule in 1917, when operations were halted in Porter County. A shortage of coal, caused by the war, kept the boats idled indefinitely. The crews were released to work on local farms.[66] The war hindered ditch work in other ways. Construction on the Singleton Ditch, meant to drain marshland in Lake County, was halted earlier in the year because landowners wanted to divert funds to buy Liberty Bonds.[67]

When the Marble–Powers Ditch was finally completed in March 1921, the Kankakee had been dredged from its headwaters to Water Valley in Lake County, about seven miles from the state line. As early as 1913, L. R. Williams petitioned for the final section of the dredging of the river, to begin where the Marble Ditch was to end. Like so many other ditch projects, the Williams Ditch was in and out of court until 1918, but this final section of the straightened channel, the Williams Ditch, was completed in June 1923, to the border of Indiana and Illinois.[68] The Place Ditch, first among ditches to straighten the river, was begun in the summer of 1901. Just twenty-two years later, over 240 miles of winding river was dredged and straightened (some of it twice), and the four hundred thousand-acre Kankakee Marsh effectively ceased to exist. Landowners had spent about $1,263,000 to straighten the river and drain the marsh, no doubt a large sum in 1923.[69] This works out to just over $3.00 an acre, a bargain in any year.

This consummation marked the end of a project that John Walker had only dreamed of in 1853, that the Kankakee Valley Draining Company had bungled in 1872, that the Board of Commissioners for the Removal of the Limestone Ledge pursued through setback after setback, and that the Indiana General Assembly had consistently refused to support. In spite of that lack of support, Dixon Place, Charles Tuesburg, Horace Marble, W. F. Powers, and many, many others had believed so strongly in their cause that they braved bitter attacks from their neighbors in and out of court for over twenty years. In 1872, William Hannah and some who supported the Kankakee Valley Draining Company had been excoriated in the press and in the legislature, and driven from their homes. W. H. H. Coffin, superintendent on two different dredging projects, had received death threats, which apparently was not unusual.[70]

What accounts for this tenacity? Landowners and county courts had stopped the project cold in 1872, and the legislature abandoned it until 1893, when it adopted a program to remove the rock ledge, with questionable results. After that, it refused any requests for funding the draining of the marsh. The straightening of the Kankakee River never had the full support of landowners or county courts even after 1896, when men with large landholdings revived the project, and pushed it doggedly to success.

Straightening the Kankakee succeeded in spite of sustained, and at times bitter opposition. Part of the reason for this success was the determination of men with a vision. John Walker, William Hannah, Dixon Place, Charles Tuesburg, and others believed in draining the marsh not only as a way to riches, but as a *cause*. William Whitten touched on this, wittingly or otherwise, when he linked draining the marsh to Chicago's "prodigious development," which he called its "destiny." In these terms, it was the marsh's destiny also to be developed (drained), and the men who pushed for it were the agents of that destiny, a destiny exactly consistent with the market economy that drove unrestrained exploitation of natural resources all over the Midwest, indeed, all over the country.

Landowners such as Place and Tuesburg were not only petitioners for ditching, they were strong, able advocates for the total destruction of the marsh. They succeeded where George Cass and William Hannah had failed because they traveled throughout the Kankakee Valley mustering support for the great enterprise of straightening the river. They built social and business networks across counties. They bought land— lots of it—and they drained it with open ditches and drain tiles. They believed ardently in the cause of drainage, and encomia for that cause poured from the newspapers. The *Starke County Republican* wrote that on a farm whose land was drained by the Place Ditch, "one can see forests of corn waving and tossing as far as the eye can reach in all directions. If the season is propitious for another month, the gathering of this mighty crop will prove far beyond the ability of three times the number of people the whole township contains."[71] In writing about Charles Tuesburg's syndicate in LaPorte County, *The Indianapolis Journal* gushed, "the crop of corn which the syndicate will harvest this fall will exceed 300,000 bushels, which will make 370 carloads, or more than five big trains."[72]

Landowners and newspapers were not the only evangelists for drainage. Engineers such as John Campbell, William Whitten, C. G. Elliot, and M. H. Downey, and professors like Arthur Goss from Purdue University spoke at public meetings at which they not only extolled the advantages of a drained marsh, they also offered expert testimony that draining on the scale required by the Great Marsh was *possible*. These potent voices for change never convinced all the landowners in the valley. They didn't need to. They convinced enough, and that's what mattered.

As dredge boats on the Williams Ditch worked toward Illinois in 1921, the Indiana legislature acted on a task that many thought was essential to the final success of draining the marsh. This was the dredging of the river between the state line and Momence, Illinois. William Whitten had pointed out in 1895, after some of the limestone ledge in the Illinois section of the Kankakee River had been destroyed, that more work should be done on the channel near Momence. Whitten had discovered deposits of glacial drift upriver from the rock ledge. He reported that unless the channel was "improved" [dredged] between the state line and Momence, and these deposits removed, there could be no full benefit of drainage to lands in Indiana.[73] In their report on Kankakee drainage in 1909, Downey and Pence had echoed Whitten's claim, asserting that the "full measure of success" in draining the marsh could only be achieved by dredging into Illinois.[74]

The Indiana government took these claims seriously. The General Assembly authorized the governor to appoint a three-member commission to meet with Illinois to

> [c]onsider jointly and decide upon a plan for the betterment of the Kankakee drainage system, and especially the cost and practicability of deepening and straightening the river from the Indiana state line to a point at or near Momence, Illinois, enough to take up and provide sufficient fall for the drainage work already done or in the course of construction in the state of Indiana.

For once, the legislature was willing to allocate funds for Kankakee dredging. One of the goals of the commission was to discuss with Illinois the cost of the work and to arrive at a "proportionate" amount for each state.[75]

That same month (February 1921), the Illinois legislature passed a similar resolution, but its language was cautious, only creating a commission whose task was to confer with the Indiana commission and to "investigate the conditions and proposed improvement in the Kankakee River." The commission was to report back with recommendations.[76]

The formation of these commissions was widely reported in the press, and reaction from Illinois residents was swift. In March, citizens from Kankakee County filed a petition with the Illinois commissioners, asking that the "river be unmolested in its course across the county," because of the many resorts, summer cottages, and fishing shacks on the Kankakee, and pointing out that the river "is one of the most popular outing places in the state." Residents wanted to know "why their river should be disturbed" when the Illinois side of the river would not benefit by improvement of the riverbed.[77]

In June, commissions from both states met at Crown Point, in Lake County, Indiana, and the Illinois delegation reported that "over a thousand wealthy sportsmen" who had hunting lodges and summer homes on the Illinois Kankakee opposed the dredging of the river near Momence. This group was apparently lobbying the Illinois General Assembly, and their opposition seriously threatened approval of tampering with the river in Illinois.[78]

By the fall of 1922, when dredging was very close to the state line, and no official decision about the fate of the river in Illinois had been announced, the town council of Momence grew nervous, believing that the dredge boats might be headed their way. The mayor organized a committee and appealed to W. L. Sackett, chairman of the state waterways commission, to see what the state was doing to stop "the outrage" of dredge boats crossing into Illinois. Sackett informed the men that his commission had "their eye on the state of Indiana," and that he would serve an injunction to prevent them from coming into Illinois.[79] By that time, the government of Illinois had probably made a decision about cooperating with Indiana. In 1921, the legislature had asked the Division of Waterways to investigate whether or not increased flow from dredging in Indiana would damage landowners on the river in Illinois. The investigation showed that "considerable damage would probably result" unless there was "material improvement" of the channel between Momence and the state line. This would mean great expense for Illinois landowners, unless paid for by "those in Indiana benefitting from the construction."[80]

No public announcement was ever made by either state that the two would share the cost of dredging in Illinois. Likely, Illinois was reluctant to help pay for this work, as it primarily benefited Indiana. Whatever the official reason, no dredge boats from Indiana entered Illinois.[81] No doubt, the strong opposition of the citizens of Momence and Illinois sportsmen had an effect. By 1923, the differences between Illinois and Indiana regarding the river were clear, and would remain so for the next century. Residents of Kankakee County had put the matter succinctly for Illinois: "their river" was a place of scenic beauty, for resorts and recreation. The river in Indiana was a drainage ditch, bordered for nearly eighty miles by farms and cultivated fields. Indiana had accomplished a mammoth feat of wetlands reclamation. Neither Indiana or Illinois understood the monumental consequences of this enormous environmental insult to both states. That knowledge would come later.

# NOTES

1. *Indianapolis Daily State Sentinel*, January 14, 1853. All newspaper articles, except as noted, were accessed online at Newspapers.com or Newspaperarchives.com.
2. *Indianapolis Indiana State Sentinel*, March 14, 1877.
3. William Whitten, "Kankakee Drainage," *Twelfth Annual Report to the Indiana Society of Engineers, Held at Lafayette, Ind., Jan. 12, 13, and 14, 1892* (N.p., 1892), 56.
4. John L. Campbell, *Report Upon the Improvement of the Kankakee River and the Drainage of the Marsh Lands of Indiana* (Indianapolis, 1883), 15.
5. Whitten, "Kankakee Drainage," *Twelfth Annual Report*, 56.
6. "Against Kankakee River Bill," *Indianapolis Journal*, February 4, 1895.
7. *New Albany Daily Ledger (New Albany, IN)*, December 2, 1855.
8. Whitten, *Report of Board Of Commissioners for Removal of the Limestone Ledge in the Kankakee River* (Indianapolis, 1895), 15.

9. "To Drain the Marsh," *Chicago Daily Tribune*, March 14, 1894.
10. "Levying of Kankakee Talked," *Goshen Weekly News* (Goshen, IN), January 18, 1896.
11. Reprinted in the *Delphi Journal* (Delphi, IN), October 21, 1874.
12. "Reclaiming Wet Lands," *Hagerstown Exponent* (Hagerstown, IN), August 19, 1885.
13. T. H. Ball, *Northwestern Indiana from 1800 to 1900* (Donohue and Henneberr, 1900), 441.
14. "Future of the Kankakee Region," *St. Joseph County Independent* (Walkerton, IN), August 11, 1894. All references to this newspaper and the *Walkerton Independent* are from Marilyn (Hiatt) Sherland, *The Island: Land Between the Kankakee and Pine Creek* (Walkerton Historical Society, 2007).
15. "Ditch Money Wasted," *Fort Wayne Morning Journal-Gazette* (Fort Wayne, IN), December 2, 1900.
16. William Cronon, *Nature's Metropolis: Chicago and the Great West* (W. W. Norton, 1991), 91–92.
17. Dixon W. Place, "The Kankakee Improvement," *St. Joseph County Independent*, January 1, 1897.
18. Place, "The Kankakee Improvement," *St. Joseph County Independent*.
19. "The Place Ditch," *Walkerton Independent*, April 20, 1901.
20. Rev. E. D. Daniels, *A Twentieth Century History and Biographical Record of La Porte County, Indiana* (Lewis Publishing, 1904), 78.
21. W. W. Thornton, *The Revised Statutes of Indiana: Containing all Laws of a General Character in Force September 1, 1897* (Chicago, 1897), 1215.
22. "The Place Ditch," *Walkerton Independent*, April 13, 1901.
23. C. F. Rupel, "The Kankakee Drainage Question," *St. Joseph County Independent*, January 16, 1897.
24. "Farmers will Fight the Kankakee Drainage," *St. Joseph County Independent*, January 23, 1897.
25. Place, "The Kankakee Drainage Question," *St. Joseph County Independent*, January 30, 1897.
26. Rinehart, "The Drainage Question," *St. Joseph County Independent*, February 6, 1897.
27. Place, "Mr. Place Replies," *St. Joseph County Independent*, February 13, 1897.
28. Elmer C. Price, "More About Drainage," *St. Joseph County Independent*, February 20, 1897.
29. Rinehart, "Drainage," *St. Joseph County Independent*, April 3, 1897.
30. "The Place Ditch," *Walkerton Independent*, April 13, 1901.
31. "A Monster Ditch," *Starke County Democrat* (Knox, IN), June 21, 1900.
32. "Ditch Knocked Off," *Ft. Wayne Sentinel*, June 27, 1900.
33. *Starke County Democrat*, March 14, 1901.
34. "Place Ditch is Established," *Starke County Democrat*, April 11, 1901. Beeman was later

hired by the Kankakee Reclamation Company to help secure drainage applications, according to the same paper, February 2, 1903.
35. "Place Ditch," *Wanatah Mirror*, May 2, 1901.
36. *Starke County Democrat*, July 7, 1900.
37. Daniels, *A Twentieth Century History*, 78.
38. "The Place Ditch," *Walkerton Independent*, April 13, 1901. The *Independent* is reporting the specifications listed in the advertisement for bids that month. No contemporary account that I can find provides the specifications of the ditch after it was completed.
39. Daniels, *A Twentieth Century History*, 79
40. Daniels, *A Twentieth Century History*, 79.
41. "Kankakee Reclamation Company," *Starke County Republican* (Knox, IN), December 18, 1902.
42. "Kankakee Reclamation Company," *Starke County Republican*, December 18, 1902.
43. Daniels, *A Twentieth Century History*, 80, 83. It should be noted that a great deal of this information in the *La Porte County History* was copied directly from an article appearing in the *Indianapolis Journal*, "From Swamp to Corn," October 9, 1903.
44. "Wonderful Improvements," *North Judson News* (Judson, IN), February 22, 1906.
45. "Completed," *North Judson News*, February 22, 1906.
46. "Completed," *North Judson News*, February 22, 1906, 84–85.
47. "Completed," *North Judson News*, February 22, 1906, 85.
48. "Miller Ditch Done," *South Bend Tribune* (South Bend, IN), December 2, 1904.
49. W. D. Pence and Morton Downey, *A Report Upon the Drainage of Agricultural Lands in the Kankakee Valley, Indiana*, U.S. Department of Agriculture Circular 80 (Government Printing Office, 1909), 8.
50. Pence and Downey, *Report Upon the Drainage*, 8.
51. W. V. Judson, *Report of the Board of Engineers for Rivers and Harbors*, House of Representatives Document 931, 64th Congress, 1st Session, March 20, 1916, 13.
52. "Yellow River Ditch Complete," *Knox County Republican*, November 20, 1902.
53. Pence and Downey, *Report Upon the Drainage*, 12.
54. Judson, *Report of the Board of Engineers for Rivers and Harbors*, 13.
55. Pence and Downey, *Report Upon the Drainage*, 8.
56. Pence and Downey, *Report Upon the Drainage*, 15.
57. Judson, *Report of the Board of Engineers for Rivers and Harbors*, 15.
58. Judson, *Report of the Board of Engineers for Rivers and Harbors*, 13.
59. Judson, *Report of the Board of Engineers for Rivers and Harbors*, 13.
60. *Starke County Republican*, November 12, 1923.
61. "How Indiana Farmers 'Busted' the Hay Trust," *Indianapolis News*, June 2, 1906, 25.
62. "Kankakee Drainage Decree Upheld," *Starke County Democrat*, April 15, 1914.
63. "Kankakee Drainage Decree Upheld," *Starke County Democrat*, April 15, 1914.

64. *Starke County Democrat*, November 11, 1914.
65. *North Judson News* (Judson, IN), November 26, 1914.
66. "Drainage Stopped by War," *The Lake County Times* (Hammond, IN), November 20, 1917.
67. *Syracuse Lake Wawasee Journal* (Syracuse, IN), February 7, 1917.
68. Chris Knochel, "Straightening and Deepening of the Kankakee River, June 1, 1921 to June 1, 1923," *Newcomer*, published by the Newton County Historical Society, Winter/Spring 2022.
69. Report from the Chief of Engineers on the Kankakee River, ILL and IND, Covering Navigation, Flood Control, Power Development, and Irrigation, Report to the House of Representatives, Document 784, February 26, 1931, 21. From the archives of the Kankakee River Basin Commission, Kankakee River Basin and Yellow River Basin Development Commission, Valparaiso, IN.
70. "Warned by Whitecaps," *Wanatah Mirror* (Wanatah, IN), April 4, 1902.
71. *Starke County Republican*, August 8, 1903.
72. *The Indianapolis Journal*, "From Swamp to Corn," October 10, 1903.
73. William Whitten, *Report to the Board of Commissioners for the Removal of the Limestone Ledge in the Kankakee River* (Indianapolis, 1895), 16.
74. Pence and Downey, *Report Upon the Drainage*, 15.
75. *Laws of The State of Indiana, Passed at the Seventy-Second Regular Session of the General Assembly, 1921* (Fort Wayne Printing Company), 25.
76. *Laws of the State of Illinois Enacted by the Fifty-Second General Assembly at the Regular Biennial Session* (Schnept and Barnes, 1921), 875–76.
77. "Fight River Change," *Quincy Whig Journal* (Quincy, IL), March 23, 1921.
78. "Wealthy Sportsmen Fight to Save Wilds," *The Lake County Times* (Hammond, IN), June 4, 1921.
79. *The Momence Progress* (Momence, IL), October 27, 1922.
80. *Fourth Annual Report of the Department of Public Works and Buildings, Division of Waterways, July 1, 1920 to June 30, 1921* (N.p., 1921), 30.
81. A story persists that a group of armed Illinois "river men" confronted the boats at the state line and forced them to stop before entering Illinois. This story is found in Elizabeth Morrison, *Memories of Momence* (N.p., 1976), 44. Her source is from an unpublished article, "The Kankakee, Wonderful Waters," by Neil Metcalf. It is hard to believe this really happened. Morrison gives the date for this as 1917, and the dredge boats were not close to the state line until 1922. If Indiana had tried to send dredge boats into Illinois without permission, this would have been tantamount to an invasion and is extremely unlikely. Also, I cannot find corroboration for this story from any contemporary newspaper in Illinois or Indiana, or from the Illinois Office of Department of Water Resources, formerly the Division of Waterways.

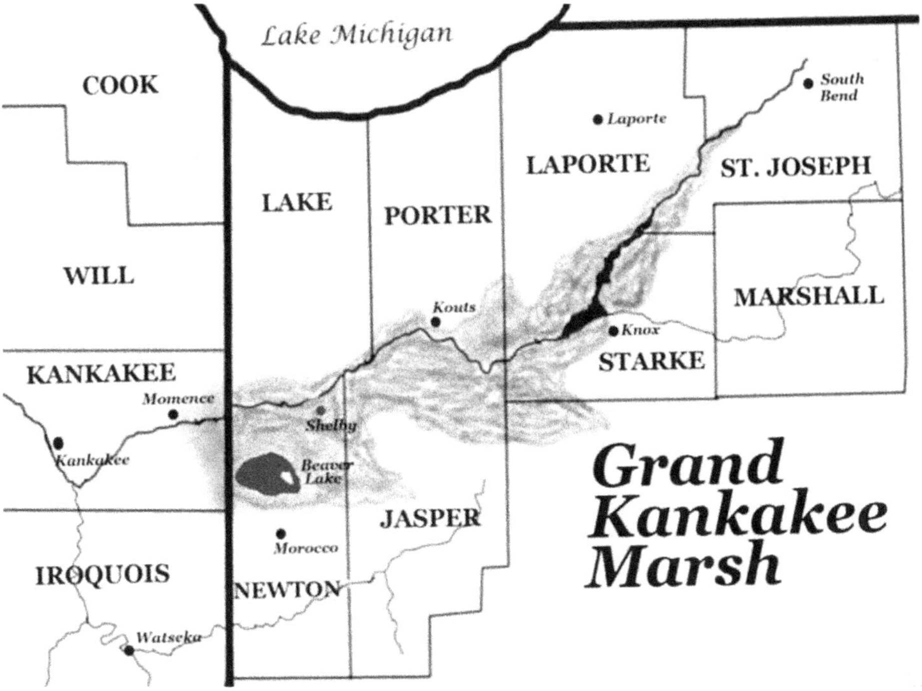

The Kankakee Marsh and the Kankakee River before draining, showing their relation to Beaver Lake and Illinois, and the eight bordering counties in Indiana. The Yellow River (unmarked) joins the Kankakee River just north of Knox. The Iroquois River (unmarked) begins in Indiana, flows southwest to Watseka, Illinois, then flows north to join the Kankakee River south of Kankakee, Illinois. Picture courtesy of the Kankakee County Museum Archival Collection.

Aerial view of the Kankakee River looking east from Illinois. The bridge marks the boundary between Illinois and Indiana. Note the unmolested meander on the Illinois side of the river. This figure appears in color online. Cover photo by Brian Kallies from the documentary *Everglades of the North: The Story of the Grand Kankakee Marsh*. Used with permission from For Goodness Sake Productions.

The sign describing Beaver Lake installed on US 41 by the Newton County Historical Society. The drab, uncultivated field in the background used to be the heart of a thirty-six-square-mile lake. The sign is big, at least ten feet tall, and dominated by an image of the lake taken from the 1834 US survey of the area. The sign's size evokes the lake's huge area, but it also makes clear the desire that the lake, now long gone, be commemorated, its absence pondered. This figure appears in color online. Photo by the author.

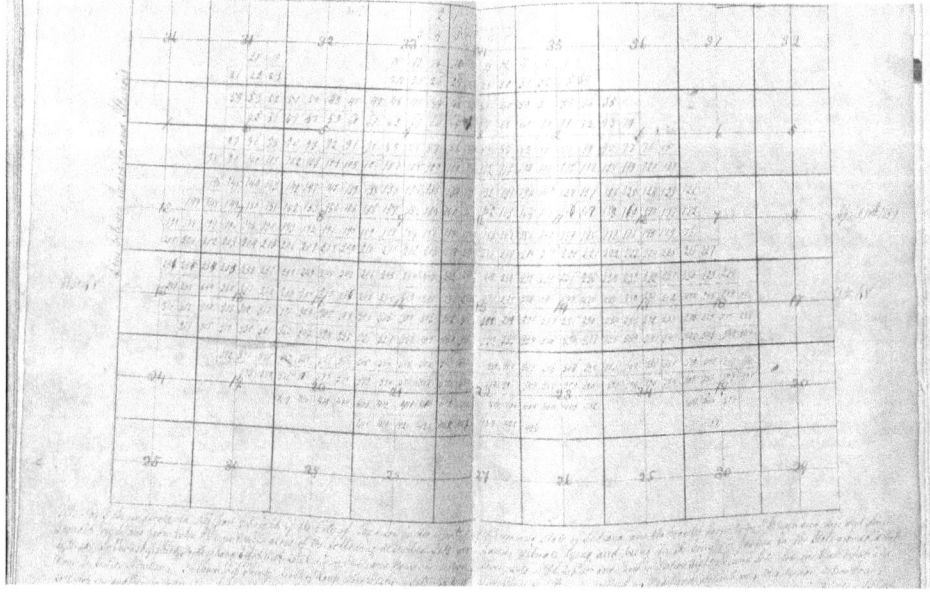

Michael Bright's plat deed of Beaver Lake, 1857, as described in Chapter 2, "Selling Beaver Lake." Note the outline of the lake surrounding 427 numbered lots, all provided solely by Bright. The writing underneath is possibly in Bright's hand. This figure appears in color online. This is a photo by the author of the deed as it appears in records in the Newton County Recorder's office in Kentland, Indiana.

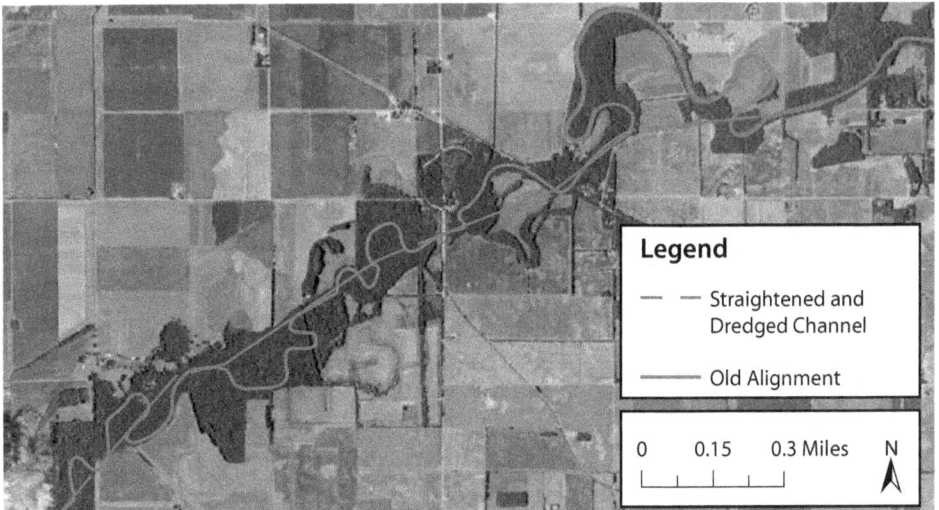

An aerial photo showing the method of straightening used to channelize the Kankakee River. The solid line represents the original, winding channel of the river. The dashed line is the new, straightened channel, dredged in some locations, and straightened in others to cut off meanders in order to create a drainage ditch. The photo shows a portion of the river in LaPorte County. This figure appears in color online. From "The Future of the Kankakee Basin," courtesy of Scott Pelath, executive director of the Kankakee River Basin and Yellow River Basin Development Commission.

One of the floating dredge boats used to dredge and channelize the Kankakee River. Dredges like this turned over two hundred miles of winding river into a an eighty-mile drainage ditch in just over twenty years. Nelson, "Dredge on the Kankakee - Burrows Camp," ca. 1909; digital image, Steve Shook Postcard Collection; Northwest Indiana Genealogical Society. Used with permission.

TYPICAL TENANT HOUSE, KANKAKEE RE-CLAIMED LAND

A typical tenant farmstead built by Charles Tuesburg in LaPorte County, ca. 1902, on the drained marsh. From Rev. E. D. Daniels, *A Twentieth Century and Biographical Record of Laporte County, Indiana* (Lewis Publishing, 1904), 81.

The Kankakee River flood of 2018. This could be a photo of many such floods. This figure appears in color online. Courtesy of Scott Pelath, executive director of the Kankakee River Basin and Yellow River Basin Development Commission.

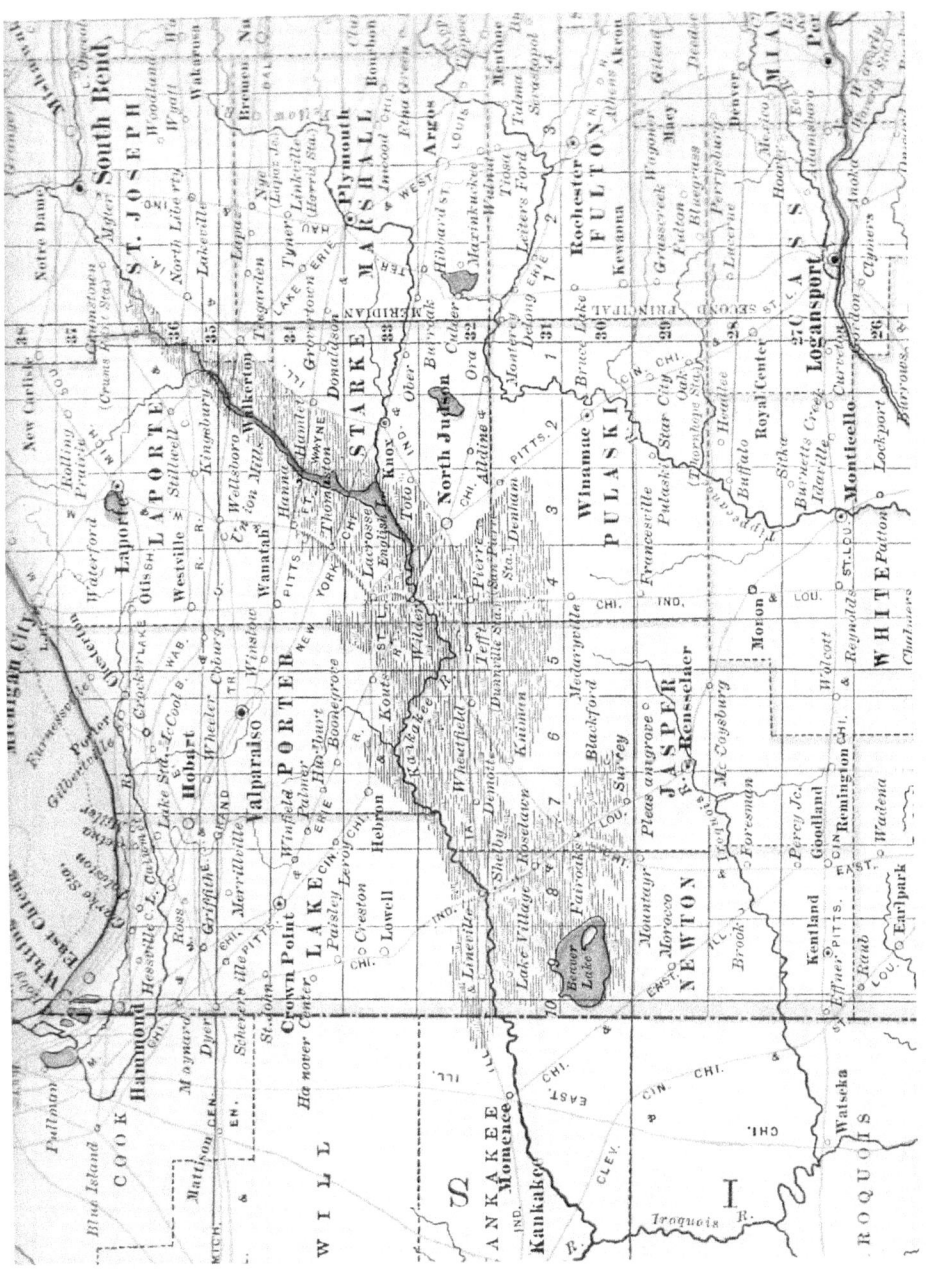

Railroads appeared in northwest Indiana in the late 1850s. By 1897, a number of them crossed the marsh, which was not yet drained. Trains were carrying farm products from northwest Indiana to Chicago years before the marsh was destroyed. Cities and towns that appear on this map exist today. Railroads provided access to the river for equipment and supplies used in the dredging projects that drained the marsh in 1901–1923. Beaver Lake is misrepresented on this map; the location is accurate, but the lake was probably completely gone by 1897. Detail from *The Grand Kankakee Marsh* (The Century Atlas Map Co., 1897).

**Izaak Walton League Monthly**
**Defender of Americas Out-of-Doors**
**Official Organ I.W.L.A.**

Volume I        AUGUST, 1922        Number 1

## TIME TO CALL A HALT

### By EMERSON HOUGH

IN this year, 1922, the lovers of outdoor America for the first time began seriously to realize that outdoor sport in this country soon will be a thing of the past.

Scrambling for the last remnants of our great heritage, we have been so busy as to be blind. Now the truth comes home. Now for the first time, a sudden consternation comes to the soul of every thinking man who ever has loved this America of ours.

**It is time to call a halt.** There is not left one honest, disinterested, unselfish agency devoted to the preservation of outdoor America. Of the great bureaus of our National government, the National Park Service, the Forest Service, the Biological Survey, there is not one which has not proved itself an agency of destruction and not of preservation of outdoor America. With them, always the record shows the bureau first, America last. **It is time to call a halt.**

Of these journals ostensibly devoted to the preservation of outdoor America, there is not one that does not show itself devoted to commercial gain; not one which, for that reason, is not rather an agency of destruction than of preservation of outdoor America. **It is time to call a halt.**

Of the alleged protective leagues there is not one which does not have commercial or personal gain or aggrandizement under it as its real basis, which is not rather an agency of destruction than of preservation. **It is time to call a halt.**

Of the alleged true sportsmen of this country, those who use rod and gun, not ten percent have practiced the creed which hypocritically they profess. Claiming self denial, we practice self indulgence. Which shall first cast a stone? And yet, my brothers, **it is time to call a halt.**

Never has transportation been so cheap, so rapid. There is no longer any wilderness. Betrayed by its guardians, forgotten by its friends, the old America is gone and gone forever. Never again shall we have more than fragments. If even these be dear, **THEN SURELY IT IS TIME TO CALL A HALT.**

These are not harsh words, or thoughtless words, or bitter. They are only unwelcome words. They are unwelcome because they are true. But no man ever gained anything by deceiving himself. We have been doing that. **It is time to call a halt.**

Can any human agency work the great miracle of giving the ages a part of the America that was ours? I do not know. I dare not predict.

Can this weak, new, little journal, openly established as a pulpit of heresy to the orthodox selfishness and commercialism in sport, work that vast miracle? I do not know. I dare not predict. But may we not all at least join in that clean hope? Surely, if it also shall fail, then all hope of outdoor America also has failed and failed forever.

Spirit of the Great Angler; all spirits of patriots and gentle men, look down upon us and have pity upon us! We are weak. Give us of your calm and serene strength, your eternal youth, your cleanliness of soul, your lofty aristocracy of thought. Help us set aside material motives. Help us work out the great miracle, in a land now almost beyond the aid even of miracle.

When one unclean hand touches the management of this experiment, then it fails. When one commercialized motive comes into its thought, then it fails. When it becomes the organ of any man's vanity, the tool of any man's selfishness, then it fails.

At the suspicion of any one of those things, at least one name will never again appear on any of its pages. I willingly lend it here after fifty years of love and myself no better than the next — none the less with an undiminished love for this America of ours, and a hope not yet wholly faltering that the needed miracle EVEN YET MAY COME.

*Emerson Hough*

The cover from the first edition of the *Izaak Walton League Monthly* magazine showing Emerson Hough's fiery denunciation of contemporary conservationist organizations and journals. From William Voigt Jr., *Born with Fists Doubled: Defending Outdoor America* (Izaak Walton League of America Endowment, 1992), 19. Used with permission from the Izaak Walton League of America.

## How Man Does Improve On Nature

Ding Darling Cartoon from *Izaak Walton League Monthly*. Note that the upper drawing shows marshland and a prosperous farm coexisting. The bottom drawing shows both turned into a wasteland by a drainage ditch. From William Voigt Jr., *Born with Fists Doubled: Defending Outdoor America* (Izaak Walton League of America Endowment, 1992), 41. Used with permission from the Izaak Walton League of America.

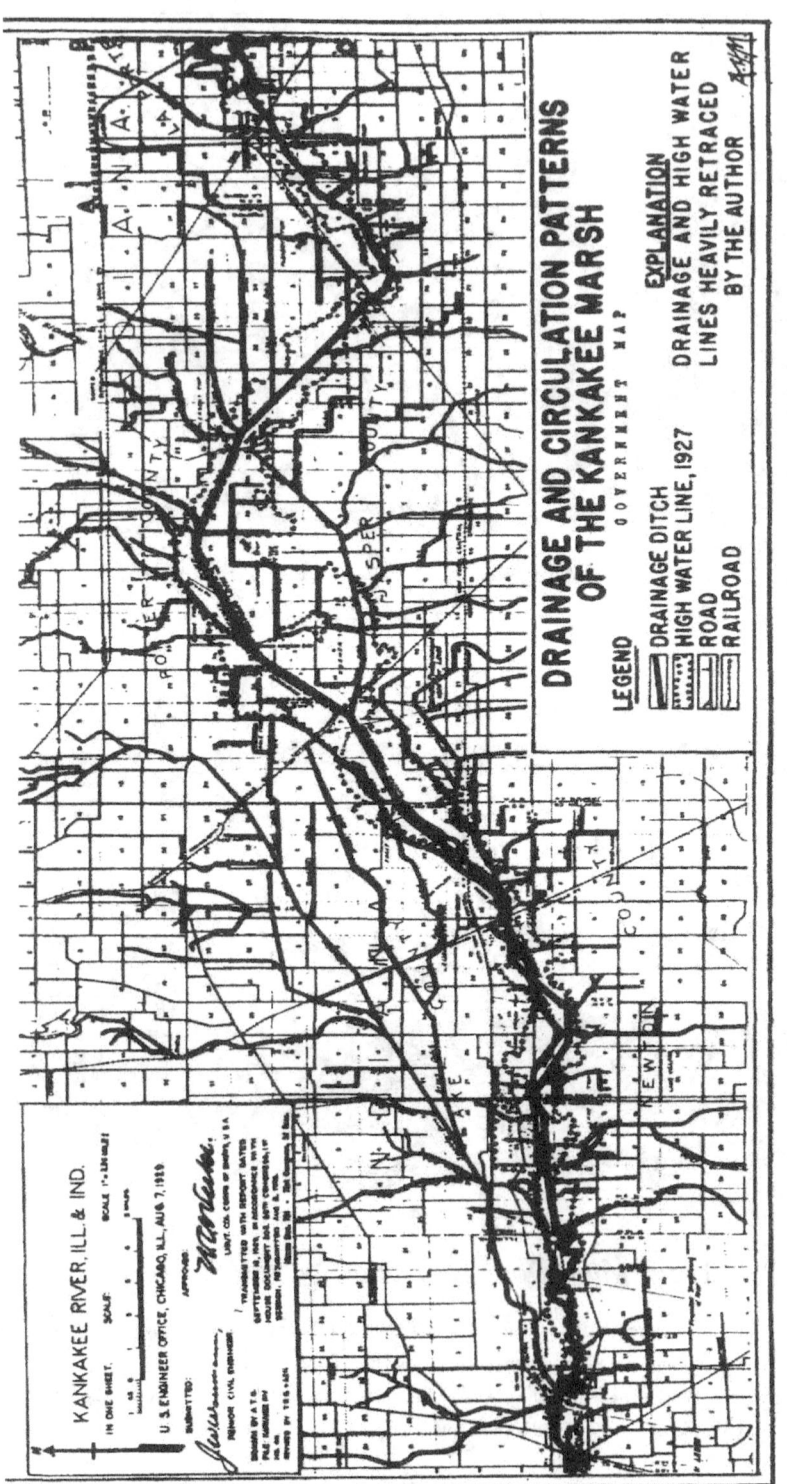

This 1929 drainage map by the US Army Corps of Engineers shows the elaborate network of lateral ditches and dredged creeks used to drain the marsh into the Kankakee River. Alfred Meyer included this map in his dissertation (1936). The heavily retraced lines of lateral drains into the straightened river show in stark relief the scale of topographical alteration necessary to convert four hundred thousand acres of marsh into cropland. Meyer, "The Kankakee 'Marsh' of Northern Indiana and Illinois" (PhD diss., University of Michigan, 1936), Papers of the Michigan Academy of Science, Arts and Letters, vol. 21.

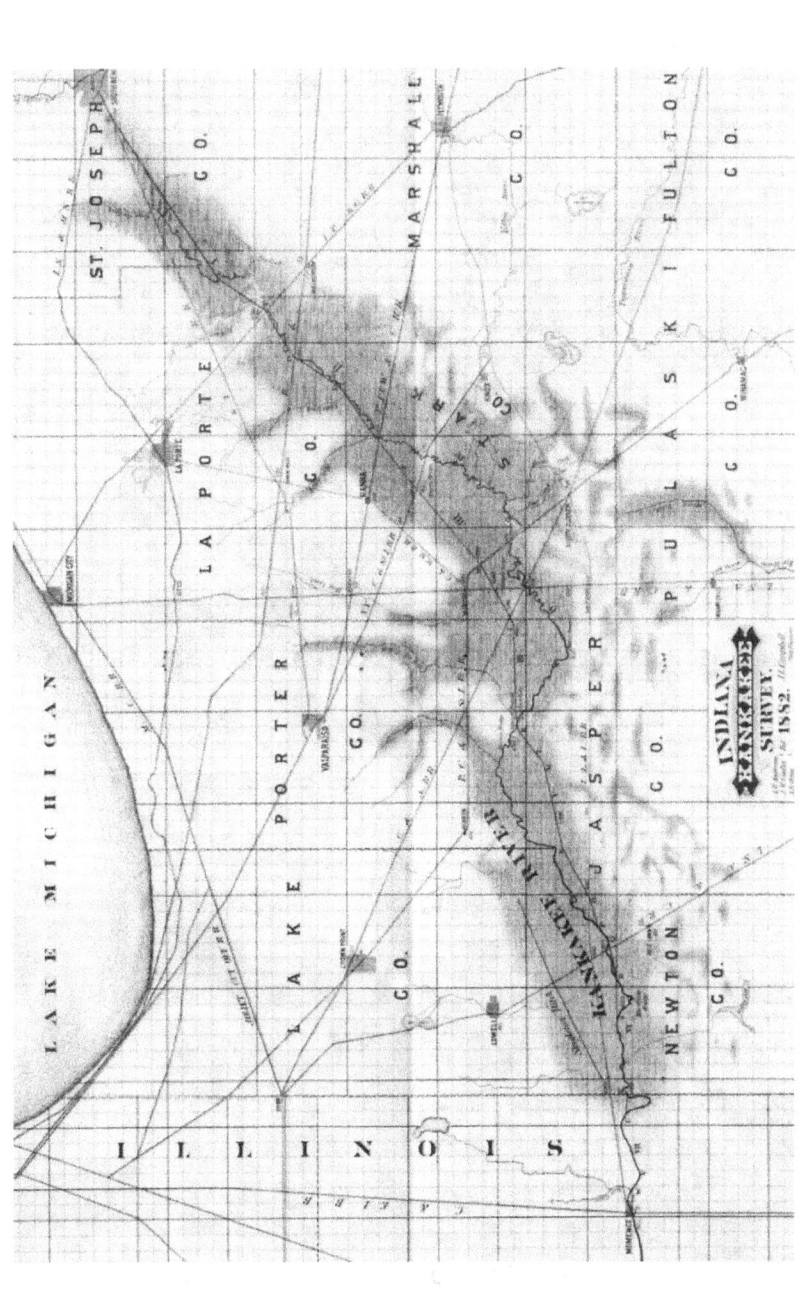

John L. Campbell's proposed rerouting of the Kankakee River, 1882 (traced by the author as the heavy black line). Note that the deviation from the original channel tracks very closely with the proposal of the Kankakee Valley Draining Company, 1871. This plan was not followed by the landowners who actually channelized the river in 1901–1923. Their method was to dredge the original channel, cutting through all meanders from the river's headwaters to the state line. Also note Beaver Lake, which by 1882 had been reduced to a small sliver. This figure appears in color online. Campbell, *Report Upon the Improvement of the Kankakee River and the Drainage of the Marsh Lands of Indiana* (Indianapolis, 1883).

Proposed rerouting of the Kankakee River by the Kankakee Valley Draining Company, 1871. This figure appears in color online. Map courtesy of the Walkerton Heritage Museum, Walkerton, Indiana.

# PART 2

# LAND REMADE: THE CONSEQUENCES

# 6

# CONSERVATIONISTS RAISE THEIR VOICES, 1925–1934

Another Brief History of Absence—and the Outcry

IN *NATURE'S METROPOLIS: CHICAGO AND THE GREAT WEST*, WILLIAM CRONON describes the profound effect that railroads, centered in the marketing hub of Chicago, had on the vast lands of the west and Midwest:

> In the second half of the nineteenth century, city and country, linked by "the wild scream of the locomotive," would together work profound transformations on the western landscape. On the farms of Illinois and Iowa, the great tallgrass prairies would give way to cornstalks and wheat fields. The white pines of the north woods would become lumber, and the forests of the Great Lakes would turn to stumps. The vast herds of bison upon which the Plains Indians had depended for much of their livelihood would die violent deaths and make room for more manageable livestock.[1]

In the third decade of the twentieth century, the Great Marsh of the Kankakee joined this melancholy list of ecosystems changed utterly by the unprecedented power of the industrial market economy. The scale of the environmental change caused by the straightening of the river and the draining of the marsh is not only hard to put into words, it is hard to imagine. To fully conceptualize the complete transformation of nearly half a million acres that included over forty miles of timbered swamp, countless sandy, wooded knolls, vast fields of marsh grass, sedges, wild rice, and countless varieties of other plant life is difficult. The size of this piece of territory was huge: eight to ten miles wide, north and south, and over eighty miles long, east to west. The marsh proper occupied land in eight different counties, and its river in Indiana drained a watershed that included over a million acres. Aside from the size, it is the nature of the utter *metamorphosis* of this territory, from one kind of landform to another, that is staggering.

In 1933, Alfred Meyer set himself precisely the task of conceptualizing, in detail, the almost unimaginable change that had befallen the marsh. In the year Meyer began his dissertation, "The Kankakee 'Marsh' of Northern Illinois and Indiana," very little was left of the original landscape. Meyer attempted to reconstruct the original marsh, and to document, in astonishing detail, the specifics of the landform of the converted marsh. The text of Meyer's work comprises an unprepossessing thirty-six pages. But in two extraordinary maps, he attempts to visualize what some geographers now would call an example of *anthropogeomorphology*, a term developed by modern researchers in Meyer's field that he would appreciate. It refers to the study of human creation of landforms that do not occur in nature, but drastically affect natural processes. The channelized Kankakee River and half a million acres of farmland, formerly a marsh, qualify as egregious examples of such human-created landforms.

Meyer called his visualization of the original marsh before it was drained, the "Fundament of the Kankakee Marsh" (Map 21).[2] The research for this map and the methodology of its construction are impressive. The map comprises twenty-eight separate township plats and several fractional plats that show the entire Kankakee Marsh in Indiana, from the Illinois border in the west to South Bend in the northeast. The township plats are the original federal plats surveyed in the 1830s, and Meyer used notes from the surveys to provide symbols showing the location of the timbered swamp along the river and the grassy marsh areas with their upland, sandy "islands" that stretched from three to five miles north and south of the river. Text accompanying the map draws further on surveyors' notes to provide details about the varieties of plant forms in each area.

The second map is remarkable, and even more impressive than the fundament map. Meyer called this the "chorography" map (Map 23); that is, it shows the details of the reclaimed marsh, its "landforms, drainage, soils, natural cover, land use, settlement, and culture."[3] This map is almost twice the size of the first, unfolding to over four feet in the printed dissertation. In his notes to this map, Meyer explains that the original was over seven feet long and almost two feet wide. This map shows the marsh region as it appeared in 1933 and attempts to show symbolically not only every soil type, but also the crops planted in representative sections of farmland. The map is incredibly detailed and cannot be easily deciphered in its reduced form. Its legend contains symbols for roadways, railroads, towns, schools, and farmsteads. There are locations of churches, cemeteries, recreation areas, and state game preserves. Even shanties and abandoned houses are shown. Meyer comments that farmland on his original map (the seven-foot map) was differentiated by color for each crop grown: corn, wheat, oats, rye, barley, buckwheat, timothy, clover, alfalfa, soybeans, onions, and mint. Areas of pasture had their own color. Such a map must have been a huge mosaic of colors, interspersed liberally with symbols marking locations of typical farmland habitations.

Meyer wanted the chorography map "to convey an instantaneous and integrated impression of the region—a unified picture of the landscape."[4] The impression conveyed is of a human-constructed built environment made hugely complex, its complexities shown in rich detail only because *they could be known*; that is, the map-maker has at his disposal soil maps, road maps, railroad maps, land surveys, topographical maps, estimates of crop acreage, and personal observation of existing structures. The map of the original marsh (Map 21) is much smaller because details of its characteristics cannot be known, only approximated—"best guesses" based on personal observation of relict vegetation, and a few eye-witness accounts from former times. Meyer supplies information about vegetation only in notes to this map; he does not locate specific kinds of vegetation in specific areas symbolically as he does with crops on the second map. For example, Meyer points out in a note that

> wild rice, once the harvest-haven of wild fowl, was found by the author in only one locality—a drainage ditch, while marsh hay appears as a relict formation in small and widely scattered, poorly drained strips or spots, generally pastured, which the dredge and the plow have not yet invaded.[5]

Meyer lists five different "associations" of vegetation in his notes to this map, such as maple–ash–elm, and pin oak–black oak, and says that these should "aid the reader in reconstructing the early formations," but it is hard to reconstruct formations when their locations are not indicated on the map. Almost all of that vegetation had been completely destroyed. Meyer suspected that it had been there, but its location had been obliterated. He comments that in the swampy areas along the river

> the primeval wilderness of virgin timber with its profusion of animal life, has all but disappeared, leaving only a semblance of its former glory in those tracts, which, though exploited for their marketable timber and firewood, have been spared the ravages of fire and disease. Extensive areas affected by the latter form of clearing present the most pitiable spectacle of despoilation—a veritable "No-Man's Land."[6]

The impression conveyed by these extraordinary maps is bleak. They suggest one landform was annihilated by another, as in a war. (Meyer's reference to "no-man's land" brings to mind World War I, a matter of recent history for him.) One cannot help but think that Meyer saw it that way, too. The "primeval wilderness of virgin timber" he refers to contained "trees of extraordinary size, 2 to 3 and even 4 feet in diameter." Meyer quotes the Indiana Commissioner of Fish and Game who described the swamp forest and its wildlife in the lower reaches of the Kankakee River only a few years before it was destroyed:

> The river courses through a forest of great elms, and trees such as grow on wet ground, and the branches of these hang near, or touch the water. Some of the trees are beautifully hung with vines, and all in all the scenery is altogether unlike that of any

other river in Indiana.... Beautiful birds flush at every turn, kingfishers, green herons, great blue herons, wood ducks, mallards, hawks, owls, occasionally a great eagle, many red heads, flickers, red wings, divers, mudhens—all these and others are there in plenty... Fish disturb the water. Set your hooks for them. Some of them are carp, but not all; many are catfish up to 15 pounds, walleyed pike of like weight, ordinary pike, or pickerel of weight sometimes twice as great, and bass, large and small mouths, up to 5–6 pounds.[7]

All this—and much more—was lost. If we add waterfowl that frequented the marsh, we might note the disappearance of whooping cranes, trumpeter swans, and the vast majority of sandhill cranes. Ducks such as the American wigeon, redhead, green-winged teal, and shorebirds such as Wilson's phalarope, lesser yellowlegs, and the solitary sandpiper, once plentiful on the marsh, were gone. This is also the case with the horned grebe and common loon. Some bird species, common on the marsh, were completely extirpated from the state, such as the Carolina parakeet, the ivory-billed woodpecker, the swallow-tailed kite, and the common raven.[8] Mammals that had thrived in the marsh, such as beaver, otter, and mink were either gone or had become extremely rare even before the marsh was completely drained.[9] The beaver were extirpated from the state by 1900, and had to be reintroduced from Wisconsin and Michigan. Wetland drainage, including drainage of the Great Marsh, caused drastic reductions or outright disappearance of some species of fish populations, such as alligator gar, bantam sunfish, and cypress darter.[10] Loss of the marsh meant the eventual extirpation of many wetland plant species such as northern willow herb, dense cotton grass, red manna grass, least duckweed, and many others.[11]

Meyer wanted his work on the marsh to be a guide to the redemption of some of this destruction. He became interested in the Kankakee region because of the work of the Izaak Walton League in promoting restoration of the marsh, and in the "Explanation" to Map 21, the map of the original marsh, he writes,

> This map and notations reveal the plant-restoration potentialities in the way of food and shelter for various types of game. The companion map, the chorographic map, supplies data on culture incident to the reclamation program in addition to the existing natural forms of the environment, all of which are significantly involved in any program contemplating partial marsh restoration or other changes in land utilization.[12]

Meyer had his eye on current events. The conservation-conscious Franklin Roosevelt administration had created public work projects in the early 1930s designed to put unemployed men to work restoring natural areas. Calls for restoration of the marsh began even earlier in Indiana. In 1925, only two years after the river was channelized to the state line, condemnations of draining began to appear. In that year, the *Brazil Daily Times* published an article headlined "Drainage of Kankakee Swamp Serious Mistake." The article is short, only five brief paragraphs, and quotes no specific source. Datelined "Indianapolis," it begins: "One of the most notable mistakes in drainage operations in the history of Indiana is seen by conservation experts in the draining of the Kankakee River marsh in the northwestern part of the state."[13] The unnamed "conservation experts" claimed that "much of the drained area" was not the productive farmland that had been promised. These "experts" were likely members of the Izaak Walton League, established in Chicago only a few years before in 1922. That the draining of the marsh resulted in little productive farmland was a recurring theme in the Izaak Walton League literature of the time.[14]

Conservation—at least of fish and game—had become very popular in Indiana.[15] Attitudes toward draining were changing—and so was the language used to describe it. In 1929, an announcement about pending restoration in the marsh appeared in the Valparaiso *Vidette-Messenger*. The article referred to the draining of the marsh as the "Rape of the Kankakee," and said that the Indiana Department of Conservation was planning a fish and game preserve in the region in order to restore "the ancient forest paradise which was destroyed."[16] The change in attitude toward the marsh expressed in this description is head-spinning. For half a century, newspapers had heralded the dredging and straightening of the river as "improvements" that would "reclaim" a "useless swamp" that had "injured the land." Now the marsh—whose literal transformation into farmland had been completed only six years previously—had become, rhetorically, an "ancient forest paradise." (The *Vidette-Messenger* was probably referring to language used by members of the Izaak Walton League, as discussed later.) The project that Dixon Place had called a "blessing for future generations" was now labeled by newspapers, without attribution, not only as destructive, but as a malicious, even criminal act—as a "rape."

The *Vidette-Messenger* article was an announcement that the government of Indiana itself had begun to consider restoring some parts of the marsh. By 1929, natural areas were being preserved all over the state. The movement for this began in 1919 when the legislature passed the Conservation Act, creating the Indiana Department of Conservation (eventually the Department of Natural Resources), which had broad powers over the state's natural areas. The department could enforce all state laws regarding land and waters, fish and game, forests, and even insects. Officers of the department could investigate matters pertaining to natural resources, and could issue subpoenas and examine witnesses. The department also had the power, with the approval of the governor, to purchase land for state parks, using the right of eminent domain if necessary.[17]

The driving force behind this legislation was a single individual, Richard Lieber, a well-connected businessman from Indianapolis. Lieber had become interested in conservation after a visit to Yosemite Park, and had attended President Theodore Roosevelt's White House Governors' Conference on Conservation in 1908. He had long advocated for the creation of a state park system, and strongly supported the establishment of a department of conservation. He served as director of the new department when the legislature finally created it in 1919.[18]

One of Lieber's new preserves, announced by the Valparaiso *Vidette-Messenger* in March 1929, was a project to restore part of the marsh. The story wildly exaggerated the size of the area at twenty-five thousand acres—the office of the Department of Conservation actually purchased only 3,556 acres in Jasper and Pulaski counties in December, with the expectation that five thousand more acres would be acquired later. The department intended to reflood the area to "produce wild duck marshes and to make the place a haven for wild waterfowl, and a sanctuary" for wildlife. The preserve was named the Jasper–Pulaski State Game Reservation, and was located on what the state called "waste lands"—lands that had been drained, but were useless for agriculture.[19] Jasper–Pulaski was one of the earliest conservation set-asides in Indiana, the first preserve in the marsh region, and one of the most important to the restoration of wildfowl habitat in the Midwest. Sandhill cranes, once abundant in the marsh but non-existent after draining, began to gather at Jasper–Pulaski in 1935. Within about fifty years, they numbered in the thousands, returning every spring and fall in their annual migrations to and from Canada.[20]

What had changed in the few years since the completion of the project to drain the marsh? The architects of that project, men like Dixon Place and Charles Tuesburg, and the engineers who debated the best ways for draining, such as William Whitten and John Campbell, doubtless would have been stunned to learn that the state would someday actually pay to *reflood* part of the old marsh, and reestablish wildlife habitat there. True, the Department of Conservation's policy was to locate such areas only on areas unfit for other use, but the department was investing almost $38,000 to *restore* the land to something like its original condition.[21] In other words, the state was proposing to spend money to *undo* draining that citizens of that area had paid for, and that the state had liberally enabled with over a half century of legislation.

One of the major changes was that a man like Richard Lieber, an ardent conservationist, had advocated for, and become director of, a state bureaucracy specifically dedicated to conservation—and he knew how to use it. In 1916, before the creation of the Department of Conservation, he had founded two of Indiana's first state parks with private donations. As director of the department, he oversaw the creation of ten more parks in only fourteen years and the establishment of the first nature preserve on the area of the Kankakee Marsh. Lieber brought the same determination to saving natural places that men like Dixon Place and Charles Tuesburg had used in destroying them.

Lieber could not do this alone. He had to have the cooperation of the governor and the General Assembly, and by the 1920s social and political attitudes toward the natural world were beginning to change. This was in part due to the huge influence of Theodore Roosevelt earlier in the century. In 1908, at the White House conference on conservation that Lieber attended, Roosevelt had articulated something that many Americans were beginning to experience: a sense of loss about the natural world, a sense that something vital had been removed from the national view and could not be replaced. He proclaimed that the country

> [B]egan with an unapproached heritage of forests; more than half the timber is gone.... [W]e began with soils of unexampled fertility, and we have so impoverished them by injudicious use and by failing to check erosion that their crop-producing power is diminishing instead of increasing. In a word, we have thoughtlessly, and to a large degree unnecessarily, diminished the resources upon which not only our prosperity but the prosperity of our children and our children's children must always depend.[22]

Roosevelt's conservationism, though based on a sense of diminishment of natural areas, was basically utilitarian. He was advocating for what he called the "wise use" of resources, for like John Locke, he saw nature as valuable primarily as it provided the raw material for human prosperity. Lieber was no doubt inspired by Roosevelt, but he saw something that Roosevelt and other conservationists did not emphasize. He created public parks in locations he saw as "beauty spots."[23] He felt nature must be preserved not just to be used, but to be experienced for its aesthetic appeal, that it should be valued because it was beautiful.

Roosevelt's conservationism, and Lieber's, was propelled by a group that is often overlooked by histories of American environmentalism but is vital to an understanding of efforts to restore the Kankakee Marsh: duck hunters and fishermen. As the nineteenth century drew to a close, these individuals saw something disappearing that was more profound than the loss of fish stocks and game. This was the sense of beauty and awe that they experienced as outdoorsmen, what Judge Hunter was trying to express to Burt Burroughs (quoted in chapter 1) when he described hunting waterfowl on Beaver Lake:

> Man! Man! Man! The spectacle was too stupendous for words! He would had to have known something of those days; he should have lived in the swamps as I did, weeks and months at a time; he would have to have the echo of the deafening clamor of all this wild life in his ears, and envision the gray-white bodies and yellow legs of these mighty hosts all set to drop into the open water spaces among the rushes and wild rice![24]

John Reiger, in *American Sportsmen and the Origin of Conservation* argues persuasively that American sportsmen, men of the upper class who hunted and fished for pleasure and not necessity, began an environmental movement in the years immediately after the Civil War that was vital to the national impetus for conservation in the early twentieth century. By the 1870s, this campaign gained momentum when publications designed expressly for lovers of the outdoors provided a means for creating a sense of community and group identity:

> With the establishment in the early 1870s of national newspapers such as *American Sportsman, Forest and Stream*, and *Field and Stream*, sport hunters and fishermen acquired a means of communicating with each other, and a rapid growth of group identity was the result. Increasingly, they looked upon themselves as members of a fraternity with a well-defined code of conduct and thinking. To obtain membership in this order of "true sportsmen," one had to practice proper etiquette in the field, give game a sporting chance, and possess an aesthetic appreciation of the whole environmental context of sport that included a commitment to its perpetuation.[25]

Reiger points out that the number of sportsman clubs and associations grew rapidly during this period, and he reminds us of the important role that these groups played in forcing state legislatures to pass laws regulating hunting. The Boone and Crockett Club, formed by Theodore Roosevelt and George Bird Grinnell in 1887, was especially active in the formation of the first national parks, forest reserves, and wildlife refuges.[26]

The nineteenth-century Kankakee Marsh was a favorite location for many such clubs from Indiana and surrounding states. The Chicago Sportsman Club, the Capitol Club of Indianapolis, the Rensselaer, and the Dally Clubs were established at Water Valley, near Shelby in Lake County. Baum's Bridge, in Porter County, was home to clubs from Pennsylvania and Kentucky. Other clubs on the Kankakee were the DeGuila, Alpine, Prairie, Wallace, La Fayette, and Crawfordsville.[27] Members of these clubs must have watched with dismay as their hunting and fishing grounds disappeared when the river was trenched by dredge boats and the marsh was drained. It should be remembered that men from such clubs on the Kankakee River in Illinois, as discussed in the previous chapter, had likely been instrumental in keeping Illinois from granting permission to Indiana to dredge the Kankakee past the state line in the early 1920s.[28]

Many of the clubs on the Indiana part of the Kankakee were owned and operated exclusively by men from other states, and in the 1870s and 1880s they were resented by local hunters and farmers.[29] However, by the early twentieth century, many in Indiana, like sportsmen in other states, were beginning to be alarmed by the rapid disappearance of natural areas. By 1925, nearly fifty chapters of the Izaak Walton League had sprung up in the state, and some 172 fish and game protective organizations had been created.³⁰ The Izaak Walton League was to play a prominent role in advocating for restoration of the Kankakee Marsh.

Founded in Chicago in 1922, the League began as a pugnacious newcomer to the burgeoning conservation movement of the time. The first edition of its monthly magazine featured a cover containing a fiery manifesto by Emerson Hough, an avid outdoorsman, novelist, and a writer for *Forest and Stream*. The headline posed a challenge that drew a rhetorical line in the sand: "Time to Call a Halt," it proclaimed in large type. Hough's article was a *J'Accuse!* of establishment conservationism, and his litany of denunciations took no prisoners:

> Scrambling for the remnants of our great heritage we have been so busy as to be blind. Now the truth comes home. Now for the first time, a sudden consternation comes to the soul of every thinking man who ever has loved this America of ours.
>
> **It is time to call a halt.** There is not left one honest, disinterested, unselfish agency devoted to preservation of outdoor America. Of the great bureaus of our National government . . . there is not one that has not proved itself the agency of destruction and preservation of outdoor America. . . . **It is time to call a halt.**
>
> Of these journals ostensibly devoted to the preservation of outdoor America, there is not one that does not show itself devoted to commercial gain; not one of which, for that reason, is not an agency of destruction rather than preservation. **It is time to call a halt.**
>
> . . .
>
> Never has transportation been so cheap, so rapid. There is no longer any wilderness. Betrayed by its guardians, forgotten by its friends, the old America is gone and gone forever. Never again will we have more than fragments. If even these be dear, **THEN SURELY IT IS TIME TO CALL A HALT.** [All emphases in the original.]³¹

Hough not only indicted the government and outdoor magazines, he placed nature organizations and sportsmen themselves in the dock, claiming that not 10 percent "have practiced the creed they hypocritically profess." As inflammatory as this language was, it seemed to get results. At the very least, it helped energize the recruiting efforts of the League's national president, Will H. Dilg, and others. Before the end of 1922, the Chicago chapter was joined by hundreds of chapters in thirty-four states, and its magazine attracted the talents of popular writers like Zane Grey and Robert H. Davis.³² Within ten years, the League had chapters in almost every state, with a total of three thousand nationwide, and one hundred thousand members, most of them in the Midwest.³³

The League was especially concerned about damage caused by the draining of wetlands. League President Dilg declared in an editorial that "we have gone drainage mad.

We are confronted by the awful fact that we have drained an area equaling the five Great Lake states and 11 million acres besides. And the tragedy is that in our ignorance less than one-third of this wholesale drainage has resulted in even indifferent farm lands."[34] The League's energetic approach to this issue had a spectacular success in 1923. In that year the League protested a drainage scheme in a marsh along the upper Mississippi River, in Wisconsin. The League decried this as the "Drainage Crime of the Century," and was successful in galvanizing a national effort to save the wetlands from destruction. In 1924, the US Congress declared the area, known as the Winneshiek Bottoms, as the Upper Mississippi Wild Life and Game Refuge.[35]

In 1933, the League turned its attention to Indiana, and adopted the restoration of the Kankakee Marsh as an official project.[36] Conservation was riding a national trend in the 1930s, thanks to the Franklin Roosevelt administration's creation of public works programs such as the Civil Works Administration and the Civilian Conservation Corps. The idea of restoring at least some of the marsh seemed to be a part of that conversation.

The Civil Works Administration indeed had its eye on the marsh. In February of 1934 the Indiana Department of Conservation announced that it was conducting a survey of the marsh as part of a Civil Works project to determine a possible location for restoration.[37] In March, the League's magazine published an article in the take-no-prisoners tone of Emerson Hough to excoriate those who drained the marsh. Written by Bob Engels of the Indiana chapter, the article blamed the "greed of a gang of well organized gangsters" for destroying the marsh:

> The draining of this vast marsh, and the destruction of this great haven for wildlife, is an outstanding example of man's rape of nature to satisfy his lust for more land and gold. As it now stands, this waste is a fit memorial to the drainage racketeers and land sharks, who have long since moved on leaving only a path of sinful destruction and waste.[38]

Engels's use of the language of sexual violence to stigmatize those who drained the marsh, and to portray the marsh itself as a helpless victim had become a common trope—even newspapers used it, as shown earlier in the case of the Valparaiso *Vidette-Messenger*. Such language slips easily into hyperbole, as Engels shows:

> Yes, the Kankakee was drained. We in Indiana got our drainage, but the poor sinners below us on the Illinois and Mississippi Rivers, now get our flood waters. We have drained our lakes and swamps, emptied our vast reservoirs, and we have upset nature's balance. The flood waters that wash down our drainage ditches to that ugly man-made ditch, and from there into the Illinois and finally into the Mississippi,

gathering force and volume in their mad rush to the Gulf of Mexico are the same waters that cause the great floods in the Illinois and Mississippi river bottoms, leaving a path of destruction and annually doing millions of dollars of damage, destroying thousands of homes and claiming the lives of hundreds of people.[39]

Engels perhaps wanted his readers to think of the Great Flood on the Mississippi River in 1927 as an annual event. That flood was catastrophic. It displaced hundreds of thousands of people and flooded over sixteen million acres from Illinois to Louisiana. The flood was the result of exceptionally heavy rains throughout 1926.[40] Engels considerably exaggerates the contribution of the dredged Kankakee River to that disaster. By the time any excess discharge from the Kankakee actually got to the Mississippi, it would be negligible. However, flooding along the Kankakee became a major issue in later years, and will be discussed in later chapters. Engels's article concluded with the observation that the Civil Works Administration was currently surveying the Kankakee Valley for a possible restoration site. He ended with the hope that the Kankakee would "receive the blessing of Federal Aid," and be "restored to its natural beauty and splendor."[41]

Hopes ran high, perhaps too wildly high, for such a project. In March, the *Hammond Times* published a story headed by an "Editor's Note" that proclaimed the imminent approval of a "gigantic restoration project" on the marsh at an estimated cost of $2,500,000. The Editor's Note claimed that this had been advanced as a number one Civil Works Administration project, and only needed President Roosevelt's approval before it began. No sources were provided to support these extravagant claims. The article that followed was one of two in a series written by William F. Collins, Director of the Indiana Izaak Walton League, to explain "what the restoration would mean to the people of Indiana." Collins wrote in breathless prose replete with colorful metaphor that was, however, conspicuously lacking in specific information about the project:

> The Kankakee marshland is coming back. Thousands of acres of sad and weary land that was "reclaimed" for agriculture at an enormous and useless expense will again be put to work as the Creator planned. In the place of this drained tundra of desolation, in place of the forlorn miles of tax delinquent farms over which have swept the winds of depression carrying the dry dust of foreclosure, forfeited drainage bonds, [and] poverty, we will have the lush verdure of the old swamp. In a few more years, the president's plan, maturing under the able direction of the US Biological Survey and aided by our own Indiana State Conservation Department under Virgil Simmons, director, will make out of the Kankakee the conservation masterpiece it once was.[42]

In his next article in the series, published the following day under the headline "Draining Land Brought No Prosperity," Collins enlarged on the canard of the marsh as a "drained tundra of desolation," declaring that the "roseate hopes of the men who drained the Kankakee died on the operating table." Collins claimed that the drained marsh failed utterly in the promise of creating good farmland. However, with restoration he promised that

> [d]ucks, geese, wild life of all kinds will appear and multiply and overflow into the surrounding country. Mother nature will tuck her petticoats of swamp vegetation into the dredge ditch wounds and again there will be a great water conservancy reservoir to prevent floods farther down in the state.[43]

Collins ended on a ringing prophetic note, stating that the "day of the destructionist is definitely over, the day of the conservationist is in the dawning."

Collins should be forgiven his loosely founded optimism, if not the forced metaphors. The 1930s were heady years for conservationists, especially the early 1930s, when both the Civil Works Administration and the Civilian Conservation Corps were putting Americans to work restoring natural areas. And given the League's success in saving the Winneshiek Bottoms ten years before, restoring the Kankakee Marsh seemed within reach. However, the grand project announced by the *Hammond Times* and the Indiana chapter of the Izaak Walton League never materialized. A survey of the Kankakee Valley in 1932 by the US Biological Survey (precursor to the US Fish and Wildlife Service) had concluded that a major restoration of the marsh was not feasible because it would "nullify the benefits of drainage."[44] Perhaps the CWA engineers conducting the 1934 survey reached a similar conclusion. In any case, the issue became moot very quickly. The Civil Works Administration itself was disbanded that same month, March 1934.[45] As far as the Kankakee Marsh was concerned, the day of the destructionist was definitely not over.

The rhetoric of Izaak Walton League members Emerson Hough, Bob Engels, and William F. Collins seems overheated. Alfred Meyer himself called the League's conservationist advocacy "propaganda," but we should remember that Meyer's study of the marsh was inspired by the Waltons.[46] The marsh region after 1923 was clearly not a "drained tundra of desolation," as Meyer's study shows. Though some drained areas were indeed unfit for agriculture, much of the region supported prosperous farms, and does to this day.

The fiery rhetoric of the Waltons was powerful and persuasive, as Meyer knew. What was particularly different about the language of Hough, Engels, and Collins was its colorful invective, its ad hominem attacks, and its hyperbole—the flagrant exaggeration to drive a point home. This is the language of combat, and no quarter is given to the nicety of rhetorical decorum usually observed by debaters. Who is to say that this bellicose posture was not warranted?

The loss of wildlife, and the disappearance of natural places where wildlife lived, was more than alarming. It signaled that something was happening that was profoundly wrong. The Waltons, like other sportsmen who hunted and fished for pleasure, were especially horrified by the carnage wrought by commercial hunters. John F. Reiger reminds us of two famous species destroyed by market killing—the bison and the passenger pigeon. The bison numbered in the millions in 1870, he points out, but "this symbol of western expansion was utterly wiped out in thirteen years. With the close of 1883, the 'thundering herds' were reduced to a pitiful remnant seemingly destined for extinction."[47]

Passenger pigeons once darkened the skies by the millions—perhaps billions. The last one died in the Cincinnati Zoo in 1914. Joel Greenberg's description of the destruction of these birds—which once darkened Indiana skies—is particularly disturbing:

> For the last century of its life, the passenger pigeon was subjected to human depredations of a magnitude and ferocity as difficult to imagine as the bird's abundance. The huge concentrations were attacked in every conceivable way: gunners by the hundreds emerged with the arrival of the flocks; birds were netted; chicks were knocked out of their nests with poles and blunt arrows; cannons were filled with scrap and discharged into the trees, clearing large swaths of nests; trees heavy with nests would be carefully felled so as to knock over others; roosting and nesting sites were deliberately set ablaze; sulfur was burned in attempts to asphyxiate squabs and brooding adults. Pigs were turned loose to feed on the maimed and dead surplus.[48]

Reiger notes that these were "merely the most famous examples of the devastation wrought by a capitalist democracy that treated wildlife resources as mere articles of commerce, available to everyone for the taking."[49] The Waltons knew how much was slipping away, and they knew it was going fast. The same commercial impulse that animated the destruction of the bison and the passenger pigeon animated the destruction of marshland. For them, regions like the Kankakee had been treated as "mere articles of commerce, pre-empted by agricultural interests"—"destructionists"—for the taking.

The Waltons never gave up on the possibility of restoring some parts of the Kankakee Marsh, but they did change rhetorical strategies. In 1942, the Indiana Department of Conservation submitted a comprehensive plan for restoration on the marsh called "The Kankakee Basin Plan." The Izaak Walton League published the department's plan in a booklet for its members called *The Old Kankakee: The Dream of 100,000 Hoosiers Can Come True*. Len Hofmann, president of the Indiana Division of the League, wrote a Foreword to the plan. Note that Hofmann's tone is much different than that of earlier Indiana Waltonians:

> Experience has demonstrated that all too often we have destroyed real and valuable resources in pursuit of other values which all too often failed to materialize. Too, changing values have given us a greater appreciation of the resources we have destroyed in the relentless march of "development." In all too many instances we failed to weigh the values to be destroyed against the expected new values to be created. Such was the case with the Kankakee. Much of the land reclaimed for agricultural pursuits proved to be good agricultural land indeed; much proved well nigh valueless for farming. But to make this discovery we destroyed one of the greatest waterfowl concentrations in the world, one of the greatest fishing areas of the Middle West, and one of the most efficient flood control reservoirs in the Upper Mississippi watershed, and we have seen thousands of tons of rich topsoil lost by erosion of both wind and water.[50]

Hofmann's tone is greatly moderated compared to Engels's and Collins's, though he makes a similar point about some drained lands being useless for farming. He also shifts the argument from invective and ad hominem attacks to an emphasis on *values*. He implies that the marsh had a value beyond the economic that was not appreciated, and was only recognized after it was destroyed. His claim is that not only the marsh was destroyed; its *value as a natural area* was destroyed. He lists those natural values in terms of waterfowl, fish, soil, and flood control.

This argument for natural value is echoed in the language of the plan itself, when it calls for restoration of parts of the Kankakee basin in order to protect waterfowl. (The idea of "natural value" is discussed more fully in chapter 11.) This protection goes beyond local concerns, the Indiana Department of Conservation argues, because of US treaty obligations under the Migratory Bird Treaty agreement with Canada and Mexico of 1918, which prohibits the hunting of migratory waterfowl. The plan argues that "[t]he indispensability of the Kankakee as a part of the plan for North America is the true estimate of its value."[51]

The Department of Conservation was asking that only "submarginal land," that is, land that was not suitable for farming, be set aside, and recommended a narrow strip of territory close to the river in the lower Kankakee Valley as the most suitable for a game preserve, as well as part of the Beaver Lake region in Newton County. The plan was comprehensive, and ambitious. The department estimated that one hundred thousand acres could be available for the Kankakee Basin Public Lands Area.[52]

Five years later, the General Assembly was ready to appropriate a quarter of a million dollars to buy land for restoration in the marsh.[53] By the end of the decade, the department, and the Izaak Walton League, got at least part of what they wanted. In 1949, the Willow Slough Fish and Wildlife Area near the Beaver Lake basin was established, and by 1952, two more areas, LaSalle Fish and Wildlife, and the Kankakee Fish and Wildlife Areas, had been created. The Kingsbury Area followed in 1965, and in 1979, Lake County created a park near Shelby called the Grand Kankakee County Park. In 1996, a Nature Conservancy preserve was created in partnership with the Indiana DNR in the Beaver Lake basin. Such was the persistence of the League and the Indiana Department of Conservation (after 1965, the Department of Natural Resources): over sixty years after the Izaak Walton League had declared restoration on the Kankakee Marsh a national priority, about 11 percent of the region was under public protection.[54]

In 1942, the Department of Conservation had wanted fully one quarter of the marsh region set aside for restoration. By the end of the century the Indiana General Assembly had approved only about 40 percent of what the Department had asked for. This may seem paltry in light of how long it took to have some parts of the marsh restored. The important victory for conservationists, however, was that they had found their collective voice, and began to articulate an idea about natural value that animated their calls for restoration. And they had created an audience—a wide audience—for their message. But ecological restoration was only part of the problem that Indiana faced with a drained marsh and the river that had been forced from its natural course. After 1923, the state struggled with a huge, and expensive, public works challenge: how to manage a damaged river that flowed too fast, carried too much sediment, and flooded often.

# NOTES

1. William Cronon, *Nature's Metropolis: Chicago and the Great West* (Norton, 1991), 93.
2. Alfred H. Meyer, "The Kankakee 'Marsh' of Northern Illinois and Indiana" (PhD diss., University of Michigan, 1936), Papers of the Michigan Academy of Science, Arts and Letters, vol. 21.
3. Meyer, "The Kankakee 'Marsh,'" Map 23.
4. Meyer, "The Kankakee 'Marsh,'" Map 23.
5. Meyer, "The Kankakee 'Marsh,'" Map 21.
6. Meyer, "The Kankakee 'Marsh,'" Map 21.
7. Meyer, "The Kankakee 'Marsh,'" Map 21.
8. John Whitaker and Charles Amlaner Jr., eds., *Habitats and Ecological Communities of Indiana, Presettlement to Present* (Indiana University Press, 2012), Kindle Edition, 38.
9. Whitaker and Amlaner Jr., eds., *Habitats and Ecological Communities*, 102.
10. Whitaker and Amlaner Jr., eds., *Habitats and Ecological Communities*, 99.
11. Whitaker and Amlaner Jr., eds., *Habitats and Ecological Communities*, 98.

12. Meyer, "The Kankakee 'Marsh' of Northern Illinois and Indiana," "Introduction," 360, and "Explanation" to Map 21.
13. "Drainage of Kankakee Swamp Serious Mistake," *Brazil Daily Times* (Brazil, IN), June 6, 1925. All newspaper articles, except as noted, were accessed online at Newspapers.com or Newspaperarchives.com.
14. "Kankakee Marsh May Be Restored," *Hammond Times* (Hammond, IN), February 5, 1934.
15. "Conservation of Game Profitable," *Tipton Daily Tribune* (Tipton, IN), July 24, 1925.
16. "25,000 Acre Preserve is State's Aim," *The Vidette-Messenger* (Valparaiso, IN), March 9, 1929, 7.
17. *Laws of the State of Indiana, Passed at the Seventy-First Regular Session of the General Assembly, 1919* (Indianapolis, 1919), 375.
18. "Richard Lieber," IN.gov, https://www.in.gov/governorhistory/mitchdaniels/3613.htm.
19. "Game Refuge Purchased in Two Counties," *Rushville Daily Republican* (Rushville, IN), December 19, 1929.
20. Joel Greenberg, *A Natural History of the Chicago Region* (University of Chicago Press, 2002), 362.
21. "Game Refuge Purchased in two Counties," *The Daily Republican* (Rushville, IN), December 19, 1929.
22. Theodore Roosevelt, "Publicizing Conservation at the White House," in *American Environmentalism: Readings in Conservation History*, 3rd ed., ed. Roderick Nash (McGraw Hill, 1990), 87.
23. "ORIGINS: the Richard Lieber Story," posted May 1, 2018, by Indiana Department of Natural Resources, YouTube, https://www.youtube.com/watch?v=JwG9rvGJIy0.
24. Burt Burroughs, *Tales of an "Old Border Town" and along the Kankakee* (Benton Review Shop, 1925), 108–10.
25. John F. Reiger, *American Sportsmen and the Origins of Conservation*, 3rd ed., revised and expanded (Oregon State University Press, 2001), 3.
26. Reiger, *American Sportsmen*, 4.
27. Meyer, "The Kankakee 'Marsh,'" 373.
28. "Wealthy Sportsmen Fight to Save Wilds," *The Lake County Times* (Hammond, IN), June 4, 1921.
29. See, for instance, the debate in the Indiana Senate, February 23, 1887, in *Brevier Legislative Reports, Volume XXIV, 1888*, about restricting hunting on wetlands, 497. *Brevier Legislative Reports Volume XXIV, 1888* (Indianapolis, 1888), 160, https://webapp1.dlib.indiana.edu/brevier/view?docId=VAA8558-24&brand=brevier.
30. "Conservation of Game Profitable," *Tipton Daily Tribune* (Tipton, IN), July 24, 1925.
31. William Voigt Jr., *Born with Fists Doubled: Defending Outdoor America* (Izaak Walton League Endowment, 1992), 19.

32. Voigt Jr., *Born with Fists Doubled*, 18–25.
33. *Izaak Walton League of America: A Century of Conservation Leadership, 1922–2022* (N.p., 2022), 19.
34. Voigt, *Born with Fists Doubled*, 37.
35. *Izaak Walton League of America*, 58–59.
36. *Izaak Walton League of America*, 80.
37. "Kankakee Marsh May Be Restored," *Hammond Times* (Hammond, IN), February 5, 1934.
38. Bob Engels, "The Unforgotten River," *The National Waltonian*, March 1934, 5.
39. Engels, "The Unforgotten River," 5.
40. Smithsonian, *National Museum of African-American History and Culture*, accessed January 7, 2024, https://nmaahc.si.edu/explore/stories/great-mississippi-river-flood-1927.
41. Bob Engels, "The Unforgotten River," 15.
42. "Kankakee Project is Advanced," *Hammond Times*, March 30, 1934.
43. "Draining Land Brought No Prosperity," *Hammond Times*, March 31, 1934.
44. "Biological Survey Reports on Kankakee Marsh," *Outdoor America*, March 1932, 25.
45. "Civil Works Administration," *National Park Service*, accessed January 7, 2024, https://www.nps.gov/articles/000/civil-works-administration.htm.
46. Note from Meyer in "Introduction," "The Kankakee 'Marsh' of Northern Illinois and Indiana."
47. Reiger, *American Sportsmen and the Origin of Conservation*, 94–95.
48. Joel Greenberg, *A Natural History of the Chicago Region* (University of Chicago Press, 2002), 348.
49. Reiger, *American Sportsmen and the Origin of Conservation*, 95.
50. Len Hofmann, "Foreword," in "The Old Kankakee: The Dream of 100,000 Hoosiers Can Come True" (Indiana Division of the Izaak Walton League, 1942), n.p.
51. Indiana Department of Conservation, "The Kankakee Basin Plan," in "The Old Kankakee: The Dream of 100,000 Hoosiers Can Come True," 9.
52. Indiana Department of Conservation, "The Kankakee Basin Plan," 15.
53. "Kankakee Restoration Gains Momentum," *Outdoor America*, December 1947, 13.
54. Dates and acreages of all marsh conservation set-asides are taken from the Indiana Department of Resources website: accessed January 15, 2024, https://www.in.gov/dnr/places-to-go/indiana-dnr-locations/.

# 7

# LESSONS IN HYDROLOGY, 1927–1977

## The Kankakee Floods, a Selective History

*The Kankakee River will be straightened... This is indeed good news, the best in fact... Fifty thousand acres will be reclaimed, [and] the disastrous floods we are now experiencing will be of the past and the annual overflows of the river will cease.*
—STARKE COUNTY REPUBLICAN, 1902

IN 1892, WILLIAM WHITTEN, CHIEF ENGINEER FOR THE COMMISSION ON THE Removal of the Rock Ledge, delivered a lengthy paper on the subject of draining the Kankakee Marsh to the Indiana Engineering Society at its annual meeting. His thirty-seven-page report covered several topics. Whitten rehearsed the unhappy story of the Kankakee Draining Company (see chapter 4), the history of legislative inaction on draining (referred to in chapter 5), and he provided a detailed update on the project to remove the limestone ledge near Momence. He also quoted at length from John L. Campbell's 1883 report, commissioned by the legislature, which laid out a plan for draining the marsh.

Whitten may have been addressing an audience of engineers, but he also wanted the ear of the General Assembly. His thesis is that the marsh could best be drained by using a "general plan," selected, controlled, and funded by the state—something he notes that the legislature had been reluctant to do. He quotes Campbell's report at length partly because Campbell also believed in a general plan—and said so in his report. The real core of Whitten's paper, however, is a critique of Campbell's plan, which called for the straightening of the river and construction of lateral ditches as the best way to drain the marsh. Whitten has in mind his own master plan, which he will provide. First, however, he must show the weaknesses of Campbell's plan.

Whitten may have felt himself to be in a delicate position. He knew Campbell well—they had worked together on the Rock Ledge Commission until Campbell's resignation, and as a professor at Wabash College and the Chief of the United States Geodetic

Survey for Indiana, Campbell was perhaps the best-known and most respected engineer in the state. His plan had the cachet of being contained in a government-sponsored report written by a top engineer, but it was also venerable. Its fundamentals had been the basis of the Kankakee Draining Company's plan in 1869, which had been substantially the same plan put forth by John C. Walker in 1861. Whitten prepares the ground for his critique carefully by quoting Campbell himself on the topic of disagreements among engineers: "Absolute results cannot be reached, and there is always room for honest difference of opinion among engineers as to the accuracy of results." Whitten wants to make clear how disagreements can lead to beneficial results:

> The fact that engineers may, and frequently do, differ honestly as to conclusions in problems of this kind makes it all the more important that enterprises of this magnitude [the draining of the marsh] be thoroughly investigated and if differences of opinions are found to exist, all important facts are the more likely to be brought out, and therefore the probabilities are increased of a correct conclusion being reached.[1]

Whitten's critique of Campbell's plan is thorough, comprehensive, and well-researched. His central point is the fallacy of channelizing the river as a means of drainage. Whitten believes that an artificial channel, cut straight through a river's winding path, would be prone to flooding. He claims that Campbell's plan would "leave the marsh dependent, for an outlet, upon a channel running nearly or quite to full banks at ordinary stages, thus leaving the soil fully saturated at all times, and *flooded* [emphasis in the original] at every slight raise in the outlet."[2]

Whitten also believes that drainage of the marsh itself would cause flooding, and cites several authorities to support this claim. He quotes from a paper written by Campbell himself showing that drainage of wetlands increases the possibility of flooding:

> "The tile drainage of the farms, the hundreds of miles of open, straight ditches, the removal of heavy undergrowth and the clearing out of the channels of the creeks have destroyed the storage reservoirs which kept up the water supply for our rivers... These surface improvements, however, have so increased the velocities of all the tributaries, from drain tile to the creek, that the carrying capacity of the river is *overtaxed* and the danger from floods is *greatly increased* [emphases in the original]."[3]

Campbell is referring to floods along the Ohio, Wabash, and White Rivers, but Whitten points out that the quotation would be "equally applicable to the Kankakee river after the reclamation of the marshes." He summarizes by pointing out that "[a]ll of these authorities agree that drainage increases the height and frequency of floods."[4]

Whitten then provides an account of the hydrological properties of the Kankakee Marsh that would not be out of place in a modern textbook on the environmental importance of wetlands:

> The general character of the Kankakee water-shed could hardly be more favorable to a uniform flow in the river than it has been left by nature. Forty per cent, of its area, containing four hundred thousand acres of open marsh, exposed to the full effect of sun and wind, with a saturated surface, or covered with water, and consequently in the most favorable condition possible to produce evaporation, and covered with a growth of vegetation to further increase evaporation; with the bank of Crooked [Kankakee] river generally higher than the marshes adjacent, thus converting the marshes into huge shallow reservoirs for impounding the drainage water from the higher portions of the water-shed, as well as the rainfall on its own surface, and holding this water until it is either evaporated by sun and wind, or slowly percolated through the soil into the river, are conditions well calculated to produce a very uniform flow into the river into which it drains.[5]

Campbell's 1883 plan for draining the marsh had included a recommendation for the construction of a new channel for the river, and dredging parts of it, which would have effectively channelized all of it.[6] This was essentially the same plan devised by the Kankakee Valley Draining Company in 1871 (see maps). Whitten quotes from a recent paper of Campbell's, which shows a considerably different opinion about artificially changing the course of a river: "The length and direction of a river are determined by natural laws and shortening will violate these laws only to create disturbance where now are repose and stability."[7] By 1892, John L. Campbell, dean of Indiana engineers and chief advocate for channelizing the Kankakee, had apparently changed his mind—not only about the Kankakee, but also about the efficacy of channelizing *any* river. Whitten notes this passage from the same paper:

> "The ordinary flow within its banks, with its graceful windings and gentle current, is the river which needs no care except to move the driftwood and beautify its banks. The straightening of this channel is vandal engineering, and in the end will result only in defeat."[8]

It is hard to reconcile the strong support for straightening and dredging in Campbell's 1883 report and his criticism of those strategies in his 1892 paper. But neither he nor Whitten is advocating for an unreclaimed marsh. For Whitten, reclamation is imperative, almost a "necessity."[9] Whitten's idea for this is a modification of a plan outlined in Campbell's paper, which calls for no change to the existing channel of a stream;

rather, Campbell calls for the construction of a second, "flood channel" whose capacity would be large enough to contain the highest known flood. Whitten supports this idea, and suggests that a system of levees near the Kankakee might be necessary to contain any overflow from the flood channel.[10]

It should be noted here that neither William Whitten nor John Campbell ever had anything to do with the actual draining of the Kankakee Marsh. In 1892, both men evidently held out some hope that reclamation would come under the control of the legislature, which would agree to a single, unified plan for the project, a plan that would not include what Campbell called "vandal engineering." That hope was vain. As discussed in chapter 5, within five years of Whitten's discussion, landowners embarked on several expensive, piecemeal, and haphazard dredging projects that channelized the *entire* Kankakee River, and most of the Yellow River. Those projects would not only destroy a vast system of wetland ecologies, but would also drastically alter the hydrology of the Kankakee River watershed in Indiana.

Perhaps the Whitten–Campbell plan, if implemented, would have mitigated the environmental damage that occurred. Who knows? The plan as outlined by Whitten contains no real specifics, and in spite of their public advocacy for such an idea at a meeting attended by dozens of Indiana engineers, no one in charge of the actual draining of the marsh seemed to give any real thought to the fact that the monumental environmental insult they were perpetrating would have monumental consequences. What is important about Whitten's critique here is its emphasis on the dangers of flooding if Campbell's plan was implemented. I have reviewed Whitten's paper in some detail to show that nineteenth-century engineers, though perhaps not ecologically aware (the word "ecology" was hardly in common use then), knew the hydrological dangers in channelizing a river, and in draining wetlands. They were good engineers.

Unfortunately, vandal engineering was the order of the day, not only in Whitten and Campbell's time, but well into the twentieth century. "Channelization" is defined as the widening and deepening of a river channel to increase capacity, along with steepening the gradient to increase flow velocity.[11] This was done to the Kankakee, along with dredging a straight channel through the direction of flow that cut off all oxbows and meanders, which also increased flow velocity. In the 150 years from the 1830s to the 1980s, channelization was particularly common in this country, when some 320,000 kilometers (nearly 199,000 miles) of American rivers were straightened in order to control floods, drain land for farming, or improve waterways for transportation. The availability of industrial innovations such as floating dredge boats and heavy earth-moving machines like bulldozers made this kind of river modification popular.[12]

Martin Doyle, in his book on American rivers, claims that channelization affected over 7 percent of stream miles in this country.[13] Doyle provides examples:

Along the lower Mississippi River, channelization was most dramatic between 1932 and 1955, when the [US Army] Corps of Engineers executed its new flood control mandate by shortening the Mississippi by 150 miles. Beyond the Mississippi, between the passage of the 1936 Flood Control Act and the Clean Water Act in 1972, over 11,000 miles of rivers were channelized as part of Corps of Engineers projects. Meanwhile the Soil Conservation Service channelized over 21,400 miles on its own, often in smaller headwater rivers... In states with lowland streams and rivers more prone to flooding, channelization was even more intense: over 26 percent, or 3,123 miles, of streams and rivers in Illinois were channelized in the mid-twentieth century.[14]

The ecological effects of channelization are drastic. For some forms of wildlife, catastrophic. In the Obion–Forked Deer river system near Memphis, Tennessee, channelization of over 241 miles resulted in the reduction of aquatic habitat by 95 percent and waterfowl hunting by 86 percent. According to Doyle, seventy years after rivers "from North Carolina to Missouri to Idaho" had been channelized, mammals and waterfowl that had used the rivers had not returned.[15]

By the 1970s, channelization was recognized as excessively destructive, and restoration efforts began in earnest. Re-meandering streams to reverse the ecological damage caused by destroying their natural courses became possible.[16] A famous example is the Kissimmee River in South Florida, a river that has a history similar to the Kankakee's. Serious flooding along the Kissimmee prompted the US Army Corps of Engineers in 1960–71 to transform a winding, 103-mile stream into a 56-mile-long ditch. Water control structures were built to control flooding, and nearby wetlands were drained. Flooding was lessened, but these changes caused the disappearance of 90 percent of waterfowl and the loss of 70 percent of bald eagle nesting territories. Oxygen depletion caused by channelization resulted in drastic changes to fish populations. Ecological damage, coupled with dramatic hydrological changes throughout the Kissimmee–Okeechobee–Everglades system, caused many to consider restoration. In 1999, the Corps of Engineers and state and federal agencies cooperated in an effort to undo the damage. It cost a billion dollars and twenty years, but restoration of a large part of the river is now a reality.[17]

All rivers are different, and are channelized for different reasons. Channelization may have mitigated flooding on the Kissimmee, but the Kankakee River was channelized specifically to drain a giant marsh, and there were, and are, hydrological consequences, exactly as William Whitten predicted.

Problems began early. The first flood on the marsh after it was drained occurred in 1927, only four years after the ditching was completed. In December, a cold snap froze part of the Kankakee near the Illinois border, backing up water for miles upstream in the dredged, narrow channel. *The Times* (Hammond, IN) reported that "the stream dwindled to a mere brooklet west of the jam," and that most of the water east of the ice gorge was impounded, "spreading over fields and marsh lands as far as the eye can reach."[18] Dikes near Schneider broke, threatening railroad bridges, and the major north-south highway in the area, US 41, flooded, making it impassable. Shelby, near the river, was threatened, as was Demotte, several miles south. A hundred families fled their homes before the high water receded, after the highway department dynamited the ice dam.[19] Had William Whitten known about the 1927 flood (he died in October of that year),[20] he might have noted that a channel, narrowed as it was by straightening and without surrounding marshes to absorb any overflow, would easily flood "at every slight raise in the outlet." He would be smug, but correct, in saying, "I told you so."

The first damaging flood produced the first of many reports about flooding on the river. Indiana appealed to the federal government for help, and in 1931, the US Army Corps of Engineers issued its first report on flood control on the Kankakee after drainage of the marsh. According to the report, about forty miles of the Kankakee Valley had flooded in 1927, nearly half the length of the river in Indiana. The flood covered an area as wide as four miles in some areas. The report noted that there was no flooding below Momence, which meant almost no flooding at all in Illinois.[21] Flooding on the Kankakee had been a routine event in the past.[22] Before the marsh was drained, farmers sometimes harvested corn by boat. But these "floods" had occurred on a marsh (a *wetland*, after all) that was thinly populated. Converting almost *all* of the marsh to farmland had made it a built environment, and destroying the wetlands removed the natural reservoir that had contained overflow (also as Whitten had predicted). In other words, converting the marsh to farmland had created a floodplain that was prone to damaging floods *precisely* because it was economically valuable. An overflowed marsh was worthless. A flooded farm field was damaged property. If it could not be planted at the proper time, the field was almost as useless as the marsh. Such was the irony of "reclamation." Floods would occur on the Kankakee nearly every year after 1927—some, like the one that year, would be especially damaging.[23]

In 1931, the Army Corps of Engineers was ready with suggestions about how to mitigate flooding. Recommendations included further enlarging the channel through the Momence ledge and deepening the channel in Indiana for a distance of about fifty-five miles. In other words, not enough of that obstinate limestone ledge had been removed in 1893, and the ditches that straightened the river from 1901–23 were too shallow to accommodate overflows, a point that William Whitten had predicted in 1892.[24] The report estimated that the cost of these improvements would be about $2.5 million, roughly twice what the original draining project had cost. However, since the benefits from such a project were "entirely local in their nature," Major General Lytle Brown, Chief of Engineers, concluded that "no Federal participation in this project would be justified."[25] Of course, this meant that Indiana, or most likely Kankakee Valley landowners, would have to bear the cost, again, of dredging the river.

Landowners did not want the expense of flood control. Indeed, they were much more concerned about drainage than about flooding, which had received very little attention, according to the report. Since most floods occurred during the winter months, planting was not particularly affected. The report points out the effectiveness of levees in preventing floods, but notes that "most farmers feel that first-class levees would cost more than they are worth and that they would rather submit to an occasional flooding during the growing season than pay for an adequate levee system."[26]

The Corps's report of 1931 was apparently written in 1929, so did not include mention of the flood of 1930, which swept away fifty feet of a dike near Shelby and flooded 750 acres.[27] Like the 1927 event, this was a winter flood, so did not damage crops or hinder planting. However, in 1939, the Corps revisited the issue of flood control on the Kankakee, and invited interested citizens to a meeting in Momence to discuss the issue.[28] In 1941 the Corps duly produced a new report, with some new recommendations. Lowering the rock ledge at Momence was still on the list, as well as enlarging and straightening the river from Momence to the state line. Re-dredging the river for fifty-five miles upstream of the state line was not mentioned. Levees were abandoned in favor of a moveable dam to maintain flow levels, and cleaning out sandbars and outlets to sloughs was now recommended. The Corps concluded, however, that the cost of these improvements far exceeded any possible benefit.[29] The report noted as "points of interest" that large quantities of sand had been deposited between Momence and the state line because of channel erosion in Indiana, and commented that siltation (deposits of clay and soil) had reduced the low-flow depth in that channel to less than a foot.[30] This was the first official notice of a problem that would cause friction between Indiana and Illinois for many years in the future.

Two especially damaging floods occurred after 1930. In 1943 the river flooded the valley from the state line to LaPorte County, a distance of about fifty miles. It was a spring flood—and expensive. The Lake County surveyor said that "much of the county's river area was an 'ocean of water.'" Ditches (constructed to drain the marsh) overflowed. Crops such as wheat, oats, hemp, and peas were heavily damaged, and delays in corn planting were expected.[31] By the fifth day of flooding, Army engineers did not expect a levee near Sumava Resorts in south Lake County to hold because of heavy rain, and water was three feet deep in some buildings.[32] In LaPorte County, the water overspread thousands of acres of farmland. The county agent predicted that many farmers would have to plant soybeans or even hay instead of their regular crop of corn. Farms as far as three miles from the river were under three feet of water, and many people were using boats to get around their property.[33]

The mother of all floods (up to then) occurred in 1954. This disaster was preceded by continuous flooding along the Kankakee from December 1949 to May 1950, which prevented crop planting in many areas.[34] In April 1950, an unusual ice storm had produced melt-water that poured into the already swollen river and "spread into lakes—five miles wide in places—in parts of Starke, Porter, Lake and Newton Counties."[35] About 250 homes were isolated in Lake and Newton Counties, and streets were under water in Lake Village, about two miles from the river.[36] Damages for both the Kankakee and Yellow River basins were estimated at $2,361,000.[37]

Only four years later, in October 1954, heavy rains produced floods throughout the Chicago area, including Chicago itself and towns in northwest Indiana.[38] The flood of 1954 was a stunning event. Between October 9 and October 12, over six inches of rain fell, causing overflows in the Chicago, Calumet, Kankakee, and Yellow Rivers.[39] Torrential rains as far east as South Bend sent water cascading down the Kankakee, sweeping away dikes and dams constructed by National Guardsmen in Lake and Porter Counties, and flooding local highways. Guardsmen retreated from some areas because they had "lost the battle to contain the river waters."[40] Urban areas suffered severe damages. In Plymouth, on the Yellow River, 231 homes, thirty-nine businesses, and three public buildings were damaged, at an estimated cost of $618,250. In Knox, also on the Yellow, rising water turned the town into a "virtual island." More than three hundred volunteers had to sandbag the sewage disposal plant to preserve it from flooding. They used bulldozers to build earthen dikes to save corn and soybean crops worth hundreds of thousands of dollars.[41] Damages in Knox were estimated at $42,450.[42]

On October 26, President Eisenhower declared five counties in northwest Indiana a major disaster area.[43] Communities and valuable farmland in those counties, which included most of the Kankakee Valley, were under water for over a month; floodwaters did not fully recede until November. According to one report, four thousand people were driven from their homes in northern Indiana.[44] Floods on the Kankakee and Yellow rivers *alone* produced a total of $5,667,000 in damages.[45]

The public demanded answers. In 1955 and in 1956, the Army Corps of Engineers held the usual open meetings in Indiana and Illinois to assure citizens that something was being done. The Indiana General Assembly authorized a special study of the problem on both the Kankakee and Yellow rivers. Reports were written. The 1954 flood produced three separate reports, in fact, from the Indiana Flood Control and Water Resources Commission. However, the legislature did not authorize enough money for a complete survey, and only a study of the Yellow River basin was completed.[46] Damage along the Yellow was especially, and continually, severe. The 1960 report from the Commission—which was a report on the status of the investigation begun in 1955—showed that the average *annual* cost of damage along the Yellow River up to that year was almost a quarter of a million dollars.[47]

Indiana sought federal funds for flood control on the Kankakee throughout the 1960s, but nothing was available to stop damaging floods in 1968 and 1973. In June 1975, a dike near Shelby failed, and seven thousand acres of cropland flooded.[48] Then came the flood of 1976. In March of that year, heavy rain caused a swollen Kankakee to break through dikes in Lake County, sending water cascading through houses in Sumava Resorts.[49] Upriver in Porter County, residents were forced from homes, and drainage ditches overflowed, flooding farmland and houses as much as a mile from the river.[50] Families elsewhere in Lake County were forced to evacuate when dikes failed, again. Indiana Congressman Floyd Fithian called for the Kankakee Valley, again, to be declared a disaster area.[51] By the time floodwaters receded, fourteen counties from South Bend to the Illinois state line had suffered $4 million in damages.[52]

Fithian had been actively working on Kankakee flooding problems since taking office in 1975. In July of that year he had met with representatives from the US Army Corps of Engineers, the Kankakee Valley Association, and the Indiana DNR. Recognizing that Illinois should be a part of the discussion, he invited a member of Illinois congressman George O'Brien's staff. Fithian scheduled similar meetings in August and October. He wanted quick, concrete action. The Army Corps of Engineers had authority under federal law to remove debris that clogged waterways, and Fithian wanted $250,000 from the Corps for cleaning the Kankakee River. County officials agreed with Fithian, and agreed to ask the Corps for help. In spite of pressure from the congressman, the Corps refused to take action.[53]

By the end of March 1976, county officials and many others, fed up with the inability of the state and federal authorities to help materially with flooding on the Kankakee River, were searching for solutions. On March 29, a local newspaper reported on a major meeting on the issue in Lake County. At a discussion that included landowners, federal officials, and Fithian, commissioners from Porter and Jasper counties voiced their frustration with what they called "the runaround" in dealing with the flooding.

"There's got to be a button we're not pushing," Jasper County Commissioner Fred Boissy said. "After seven months, we're back where we started from." Boissy and others criticized a state plan that in their view "catered to the wishes of Illinois residents," ignoring the problems of Indiana farmers. Many favored a plan of "massive channelization" that included the removal of the rock ledge in Illinois that had so bedeviled Indiana policy makers from 1890–93. The President of the Kankakee Valley Association claimed that money spent on studies of the problem had been wasted. He called the rock ledge the "cork in the bottle" that kept Kankakee flood waters in Indiana. A farmer from Newton County concurred, saying, "[w]e have to have an outlet for our river, and that outlet is in Illinois." Cletus Gillman, of the Soil Conservation Service, said flatly that his agency would not approve channelization west of the state line, or removal of the rock ledge, because it would only "pass the problem" into Illinois, where it would cause "extensive flooding." Congressman Fithian pointed out that Illinois officials were firmly against removing the rock ledge. He stated bluntly that "it does no good to rant and rave about that damn rock ledge. It's going to be there when we are all dead and gone. Short of an invasion of Illinois, there's no way to remove it." He warned that Indiana needed to work cooperatively with Illinois, and to "start dealing with realistic alternatives."[54]

Fithian and Gilman brought into sharp focus the inability of many in Indiana to deal with the central problem, a problem that had existed in the state for as long as ditching the Kankakee, and draining the marsh, had been considered. That many at this meeting saw Illinois alone as the problem only underscored that inability. A major stumbling block to effectively managing the river was the belief, held by many landowners and officials in Indiana, that somehow the Kankakee "belonged" to them. The Kankakee flowed through two states, and its huge watershed, shared by those two states, was part of an even larger watershed in many states that drained into the Mississippi basin. The Kankakee River, like all waterways, is part of a commons. Problems with a commons must be shared by all who use it. Indeed, "realistic alternatives" should be the order of the day.

However, value systems do not change easily, and alternatives, realistic or otherwise, were hard to find. Part of the problem was that no one seemed to be sure just who had jurisdiction over the Kankakee River. The Marble–Powers Ditch Corporation, headed by Fred Boissy, had authority over the river in five counties in the lower valley; an eight-county board also had some authority; and the Indiana Regional Planning Commission (NIRPC) was also involved in river management. The Indiana DNR had the authority to issue permits for any flood control work. After a year of negotiations, county and state officials had agreed on a project to clean and dredge the river, but the state of Illinois threatened a lawsuit to stop the dredging, because it would only shift flooding from Indiana into Illinois.[55]

Early in 1977, state representative Walter Roorda proposed a solution—a "realistic alternative" to the problem of overlapping jurisdictions in Indiana, at least. Roorda introduced legislation that would create the Kankakee River Basin Commission. Composed of representatives from all eight counties bordering the river, and funded by those counties, the commission would be responsible for coordinating development within the entire Kankakee River Basin. Such a commission would have the ability, unlike the five-county Marble–Powers Corporation or other organizations, to apply for federal funds for flood control in the entire Kankakee Valley. Roorda's bill granted the commission little real power—it would not have the power of eminent domain, for example, and it put a cap of $5,000 on the commission's annual budget.[56] As weak as the proposed commission was, many opposed its creation, including the John Birch Society and the Izaak Walton League, saying it would take power away from local control.[57] However, the General Assembly understood that the time had finally arrived for serious action on the problem of Kankakee flooding. Authorization of the Kankakee River Basin Commission was approved in April 1977.

The flood of 1976 was destructive. The floods that followed in 1978 and 1979 were also destructive, as were many later floods. These will be discussed in later chapters. The creation of the KRBC did not stop flooding on the Kankakee; in fact, the Commission could do little to control it. However, authorization of the KRBC was important. Fifty years after the first damaging flood on the drained marsh, the Indiana General Assembly took a crucial step forward in recognizing that a dredged and straightened river needed special management and special funding. The history of the Kankakee River in Indiana after 1977 is the history of the KRBC. It is a fraught history. The following chapters examine how the new commission sought a way to negotiate the fiercely competing demands of Kankakee basin farmers, Illinois landowners and officials, environmentalists, the US Army Corps of Engineers, a state legislature reluctant to fund projects, and always, the relentless threat of too much water flowing too fast in too small a channel.

## NOTES

1. William Whitten, "Kankakee Drainage," *Twelfth Annual Report of the Indiana Engineering Society, at its Twelfth Annual Meeting, Lafayette, Ind, Jan. 12, 13, and 14, 1892* (N.p., 1892), 73.
2. Whitten, "Kankakee Drainage," *Twelfth Annual Report*, 74.
3. Whitten, "Kankakee Drainage," *Twelfth Annual* Report, 84.
4. Whitten, "Kankakee Drainage," *Twelfth Annual Report*, 84.
5. Whitten, "Kankakee Drainage," *Twelfth Annual Report*, 84.
6. John L. Campbell, *Report Upon the Improvement of the Kankakee River and the Drainage of the Marsh Lands in Indiana* (Indianapolis, 1883), 15. Also see the 1882 map, which accompanied Campbell's report. The plan of channelization shown on this map is virtually the same as the one by the Kankakee Draining Company in 1871. See the 1871 map, published by this company.
7. Whitten, "Kankakee Drainage," *Twelfth Annual Report*, 78.
8. Whitten, "Kankakee Drainage," *Twelfth Annual Report*, 88.
9. Whitten, "Kankakee Drainage," *Twelfth Annual Report*, 55.
10. Whitten, "Kankakee Drainage," *Twelfth Annual Report*, 89–90.
11. Andrew Brookes, *Channelized Rivers: Perspectives for Environmental Management* (John Wiley and Sons, 1988), 6.
12. Brookes, *Channelized Rivers*, 8–9.
13. Martin Doyle, *The Source: How Rivers Made America and America Remade its Rivers* (W. W. Norton, 2018), 272.
14. Doyle, *The Source*, 272.
15. Doyle, *The Source*, 273.

16. Doyle, *The Source*, 275.
17. See these websites: South Florida Water Management District: Kissimmee River, https://www.sfwmd.gov/our-work/kissimmee-river; and the *National Graphic* webpage: "Deep in Florida, an 'ecological disaster' has been reversed—and wildlife is thriving," accessed January 24, 2024, https://www.nationalgeographic.com/environment/article/kissimmee-biggest-river-restoration-ever-completed.
18. "Twenty Families on the Kankakee Are Now Homeless," *The Times* (Hammond, IN), December 23, 1927. All newspaper articles, except as noted, were accessed online at Newspapers.com or Newspaperarchives.com.
19. "High Water is Passing; Many Go Back Home," *The Vidette-Messenger* (Valparaiso, IN), December 27, 1927, and *The Times* (Hammond, IN), January 3, 1928.
20. "William Monson Whitten," *Wiki Tree: Where Genealogists Collaborate*, accessed January 28, 2024, https://www.wikitree.com/wiki/Whitten-278.
21. "Report from the Chief of Engineers on the Kankakee River, Ill. and Ind., Covering Navigation, Flood Control, Power Development and Irrigation," House of Representatives Document 784, 1931, 32.
22. "Report from the Chief of Engineers," 32–35.
23. "Status of Investigation for Flood Control and Major Drainage: Kankakee and Yellow Rivers, Indiana" (Indiana Flood Control and Water Resources Commission 1960), 9. The "Status of Investigation for Flood Control" is in the Kankakee River Basin Commission Archives, Kankakee River Basin and Yellow River Basin Development Commission, Valparaiso, IN.
24. Whitten, "Kankakee Drainage," *Twelfth Annual Report*, 73. Whitten cites the specifications for depth in Campbell's 1883 report as being too shallow. The engineers who supervised the actual dredging probably used Campbell's 1883 specifications.
25. "Report from the Chief of Engineers," 4.
26. "Report from the Chief of Engineers," 63.
27. "Kankakee at its Highest Point in 12 Years," *The Vidette-Messenger*, January 22, 1930.
28. "Call Meeting on Kankakee Flood Menace," *The Vidette-Messenger*, July 22, 1939.
29. This report is quoted in the *Kankakee River Roundtable Brochure* (Community Foundation of Kankakee River Valley, and Economic Alliance of Kankakee County, December 2011), 16.
30. Qtd. in Nani G. Bhowmik et al., "Hydraulics of Flow and Sediment Transport in the Kankakee River in Illinois," *Report of Investigation 98, State of Illinois, State Water Survey Division* (N.p., 1980), 26.
31. "County Adds Up Water Damages as Rivers Fall," *The Times* (Lake County, IN), May 21, 1943.

32. "Levees May Not Hold," *The Times*, May 26, 1943.
33. "Flood Recedes in LaPorte Area," *South Bend Tribune* (South Bend, IN), May 22, 1943.
34. J. L. Perry, Chief Engineer, "Progress Report: Investigation of Kankakee and Yellow Rivers, Indiana for Flood Control and Major Drainage" (Indiana Flood Control and Water Resources Commission, January, 1957), 4.
35. "Television Aerials Only Victims of Strong Wind," *Kokomo Tribune* (Kokomo, IN), April 11, 1950.
36. "Overcoats Still in Style as Cold Weather Stays," *Terre Haute Star* (Terre Haute, IN), April 13, 1950.
37. Perry, "Progress Report," 4.
38. "Floods Peril New Areas in 2 States," *Hammond Times* (Hammond, IN), October 12, 1954.
39. "Floods Peril New Areas," *Hammond Times*, October 12, 1954.
40. "Water Poses Threat to Highway 53," *The Vidette-Messenger*, October 25, 1954.
41. "Cold Front Halts Rain in Chicago Area," *Harrisburg Daily Register* (Harrisburg, IL), October 15, 1954.
42. "Status of Investigation for Flood Control and Major Drainage," 2.
43. "Five Counties to be Given Flood Funds," *The Vidette-Messenger*, October 26, 1954.
44. "Flood Waters Drop," *Indianapolis News*, October 29, 1954.
45. Perry, "Progress Report," 4.
46. "Status of Investigation for Flood Control and Major Drainage," 2.
47. "Status of Investigation for Flood Control and Major Drainage," 5.
48. Jill and Robert May, *Spearheading Environmental Change: The Legacy of Indiana Congressman Floyd J. Fithian* (Purdue University Press, 2022), 170.
49. "Sumava Flooded," *The Times* (Munster, IN), March 7, 1976.
50. "Kankakee Floods Ease," *The Vidette-Messenger*, March 8, 1976.
51. "Seeks Disaster Status for Kankakee Valley," *The Vidette-Messenger*, March 10, 1976.
52. "Shelby Dikes Hold as River Apparently Crests," *Gary Post-Tribune* (Gary, IN), March 30, 1978.
53. Jill and Robert May, *Spearheading Environmental Change*, 171.
54. "Getting 'Runaround' on Flood Problem," *The Vidette-Messenger*, March 29, 1976.
55. "Kankakee Floods Studied," *Herald Bulletin* (Anderson, IN), September 15, 1976.
56. "Roorda Sees Solid Base for Kankakee Basin Bill," *The Vidette Messenger*, February 22, 1977.
57. "Will Kankakee Flood," *South Bend Tribune*, February 10, 1977, and "River Basin Bill Misread: Author," *The Times* (Munster, IN), March 25, 1977.

# 8

# MANAGING THE "MONUMENTAL PROBLEMS" OF A BROKEN RIVER, 1977–1978

The Kankakee River Basin Commission vs. Floods, Sand, and the State of Illinois

STATE REPRESENTATIVE WALTER ROORDA CONVENED THE FIRST MEETING of the Kankakee River Basin Commission in July 1977, only three months after the commission was approved. A. D. Luers, head of the Lake County Soil and Conservation Service, was elected chair. Other officers were elected, and a letter from the governor was read in which he welcomed the new members to the commission and pledged his support. Expectations were high. Some said that the bill creating the KRBC was the most important to come out of the 1977 legislative session.[1] Many had long felt the same exasperation that Fred Boissy had voiced in 1976 when he complained that he could not find the "right button" to push to get concrete action for flood control. As an article in the *Michiana Farmers Journal* pointed out, residents along the river were not only weary of floods, they "were even more weary of empty promises and futile attempts to get something done." They looked to the new commission to finally address the "monumental problems" of the Kankakee basin.[2]

It was a big committee. Each of the eight Kankakee River basin counties had three members: the county surveyor, a soil and water representative appointed by the governor, and a member of the drainage board, also appointed by the governor.[3] The public took a keen interest in the KRBC. In its first year of existence, the commission's activities were reported in dozens of articles from newspapers all over Indiana, especially those in the Kankakee River Valley. Every meeting was announced in advance and meticulously covered. Some meetings were attended by a hundred or more people, and attendees who spoke were quoted at length. Sometimes the same paper carried two articles reporting different versions of the same event,[4] but the fact that the commission had no real power to actually authorize projects, and no budget to pay for them, was rarely mentioned. Expectations were high, but hardly realistic.

Representative Roorda had a project ready at the first meeting. The Army Corps of Engineers was asking the US Congress to drop a flood control project for the Kankakee River because no local agency supported it, and Roorda wanted the newly-formed KRBC to fill that role. The project, proposed after the disastrous flood of 1954, called for the construction of levees along the river in three counties.⁵ The commission, however, frustrated with the Corps of Engineers, decided to proceed in a different direction. They wanted to find out what landowners and other stakeholders thought would work, and they wanted to build a consensus for it. In the summer of 1977, few could have predicted just how difficult that was going to be.

The Commission believed its first task was to make recommendations based on the most recent study of the basin that had been completed to date. The *Report on the Water and Related Land Resources of the Kankakee River Basin, Indiana* had been in the works since 1968, when county officials had requested it from the Indiana Department of Natural Resources. Completed in 1976, the report had an impressive list of sponsors. It was a joint project of the DNR, the State Planning Services Agency, the State Board of Health, and four different agencies within the US Department of Agriculture. Over 270 pages long, including twenty fold-out maps, it was the most comprehensive study of the basin's problems ever compiled. The KRBC was correct in deciding to give the report a priority in its deliberations. Its size, and the scope of its findings, clearly demonstrated the seriousness with which the state of Indiana and the federal government were taking the problems of the Kankakee. No one would have put it quite this way, but by 1977 the damaged river and its ruined marsh had entered the modern consciousness as a colossal environmental problem. Planners were beginning to see that a continuous, massive stream of public resources was going to be necessary to sustain the river and marsh in their altered conditions as a drainage canal and cropland.

The report was complex, no doubt because its authors found the problem complex. In addition to an exhaustive inventory of topographical, land use, and climate features, it identified nine different problem areas (labeled "needs for resource development"), and discussed them at length: land treatment, floodwater damage, drainage, erosion, sediment, irrigation, water use and quality, recreation, and fish and wildlife protection. The section on flood control was the longest, and most involved. The basin was divided into thirty-five different "hydrologic units," and the number of flooded acres of cropland, grassland, forest land, and "other" in each unit was given. Statistics for floods were astonishing. According to the report, some 106,000 acres (around a quarter of the old marsh) were liable to flooding. This was an area from the state line to fifty miles upstream, parts of which the river overflowed about three times a year. Average annual damages along the main stem and tributaries were about $2.6 million, mostly to farmland and farm structures.[6]

Solutions to the problem were developed by an intricate methodology that mixed and matched alternatives. Ten different "flood protection systems" were described, ranging from no work at all, to clearing and dredging some locations on the main stem or tributaries, to the construction of different kinds of levees at some locations, or a combination of levees and dredging at various locations. Possibilities for flood control were matched with proposals for wetlands and forest protection, recreation land use, fish and wildlife protection, irrigation and drainage development, and sediment and erosion remediation.

Authors of the plan knew what they were up against. Hoping to find a plan that worked for everybody, they provided five different options with varying emphases on flood control, recreation land use, and environmental protection. They were looking for consensus among federal, state, and local agencies, residents of Indiana and Illinois, and environmental groups. They didn't get it. A draft report was circulated to study participants in October 1975 and discussed at a public meeting in Valparaiso in December. Two months later, results were released. There was no consensus among agency participants. Indiana residents were in favor of flood control measures, but did not support environmental and recreation proposals. Illinois residents strongly rejected any idea of channel work in their state (one flood control idea was to attack the rock ledge, again), and were opposed to any project in Indiana that would increase flooding in Illinois. The report acknowledged that

> the degree of flood protection desired by the agricultural interests [Indiana farmers] cannot be obtained due to the necessity for channel work (including rock excavation) in Illinois, and objection by the state of Illinois to such proposed action. Furthermore, it is recognized that channel widening and deepening of such extent would result in severe adverse effects to fish and wildlife habitat.[7]

Undeterred, the authors decided on a "suggested plan" that contained a combination of all plan elements. This plan was comprehensive. It included channel work on the river and tributaries in Indiana, as well as levees for flood control, the adoption of flood plain zoning ordinances, development of parks and trails for recreational use, and protection for wetlands, woodlands, fish, and wildlife. Costs and benefits were calculated. At an all-day meeting in December 1977, the KRBC mulled over the report and the suggested plan. Congressman Floyd Fithian attended this meeting, and noted that any recommendation from the KRBC would almost certainly be based on information in the report.[8] The KRBC decided to take the suggested plan, a compromise among all five plan options, to the public. They scheduled a meeting in Hanna, near the river, for January 12, 1978.

Even before the January meeting, misinformation about the plan appeared. In a letter to the editor of the Valparaiso *Vidette-Messenger*, Winfield McFarland of Plymouth complained about the cost. He had attended the December meeting, and referred to the report as the "most comprehensive study" he had ever seen. He pointed out that the price tag for the plan was a "staggering $124 million dollars," which was true. McFarland believed, apparently, that Kankakee landowners would have to pay that cost, and said that the savings would only be $1.3 million, which was not true. The plan estimated that the average annual benefits would be over $11 million, provide thousands of man-years in temporary jobs, at least fifty-seven permanent jobs, and attract about 1.2 million recreation visits a year. What McFarland might have pointed out was that the plan was not only expensive, it was incredibly ambitious. In addition to flood remediation, the plan envisioned

> about 230,000 acres of flood-prone and wetland areas ... dedicated to agricultural, natural or recreational uses. Riparian wildlife and fishery habitat would be maintained on about 260 miles of stream. The application of conservation measures would be accelerated on about 426,000 acres of agricultural and forest land. Smaller land areas will have temporary or permanent land use changes.[9]

In other words, the plan called for a project that would allow cropland and a partially restored marsh to coexist. It never had a chance.

Officials in Illinois had already voiced their opposition to the Kankakee River Basin study. Over a year before, Roy M. West, chair of the Kankakee County (IL) Board, had written to the Indiana DNR that his group had reviewed the report and wished

> to express [their] strong opposition to the focus and findings of such report. The focus of the report is a blatant attempt to export water problems from the state of Indiana to the state of Illinois, and more particularly Kankakee County, and the findings blithely ignore the effects of the proposals outside of the State of Indiana ... [W]e cannot endorse any alternative that proposes construction, dredging or alteration of the main channel or tributaries of the Kankakee River in Illinois or Indiana without competent or real consideration of the total effect ... The effect we see of the advocated alternatives is acceleration of the rate at which the Kankakee River is dying.[10]

Roy West was serious, and thorough. He sent a copy of his letter, together with a summary of the study to conservation and water resource officials in Illinois asking them to submit their own evaluation of the study to the Indiana DNR. His statement—that proposals for flood control in the study would accelerate the rate at which the Kankakee River was "dying"—may have been hyperbole, but it showed the strong feelings in Illinois about Indiana's treatment of the Kankakee, and implied, in fact, that Indiana was responsible for the river's imminent death in their state. West warned his colleagues that Illinois was four years behind Indiana in studies of the Kankakee basin. He criticized the Indiana plan for its "fallacious cost/benefit analysis" and its "illogical tradeoff," among other things. However, he said that Illinois needed a "multiplicity" of studies to counter Indiana's "well-documented, cohesive" plan.[11]

The dispute over the Kankakee River Basin Study brought into clear focus not only the drastically opposed views of the river held by the two states, but also the bitterness behind those views. Many in Indiana blamed Illinois for hindering flood control on "their" agricultural river, and some in Illinois believed that Indiana was killing "their" river with sand pollution, caused by erosion, which would only be increased if flood control projects in Indiana were allowed to proceed.

The January meeting in Hanna included representatives from the Indiana DNR, the US Soil Conservation Service, a representative from Illinois, and reporters from at least three area newspapers. It drew a hundred and thirty Kankakee Valley residents.[12] KRBC Chair A. D. Luers began the meeting in the spirit of compromise. He cautioned that "[n]o one will be 100 percent satisfied. No one is going to get exactly what he wants." One of the officials was more pessimistic, commenting that the plan would "displease everyone equally."[13] Robert Mast, hydraulic engineer with the Soil Conservation Service presented the plan, and explained, probably in reference to the exclusion of any dredging in Illinois, that the plan was "a political alternative. We were given political restraints and the state line was one of them." He warned of lawsuits from Illinois if any flood control proposals threatened the river in that state. Farmers who were present wanted flood relief, and they wanted the river cleared and dredged to increase flow, pressing for exactly what residents in Illinois did not want—more water coming their way. Byron Wallace, executive director of the Kankakee County Regional Planning Commission, and a co-signer of Roy West's letter, was at the Hanna meeting. He warned that Illinois residents felt threatened by the plan, and echoed concerns raised by West's letter just over a year before. He said "persons in Illinois are worried about what Indiana may or may not do. We have no desire to have this kind of flooding happen to us."[14]

The January meeting showed the KRBC that it would have to modify the plan suggested by the Kankakee River Basin report. The next meeting was held in Valparaiso in March to discuss the "plan of action for 1978," which included proposals for flood control and flood preparedness, among other items, based on the suggested plan of the Kankakee Basin study.[15] Illinois official George Benda attended this meeting. He came prepared with an Illinois report, and with attitude. After the KRBC had discussed its adoption of the basin study plan, Benda, with the Illinois Institute for Environmental

Quality, delivered copies of a draft study from the Kankakee River Basin Task Force. Benda claimed the report blamed sand deposits in the Kankakee west of the state line on Indiana farming practices. He acknowledged that the two states' ideas for the river were "direct opposites," and told the commission, "[w]e don't want you to solve your state's problems by shoving them down the river at us." Commission member Fred Boissy replied that "we don't want to drown, either, from flooding in Indiana." Benda pointed out that the report opposed any snagging (removing vegetation) or dredging of the river in Indiana and stipulated that the Illinois Kankakee be retained as "wetlands and a low-flowing stream open to wildlife and recreation." He stunned the commissioners by saying, "[w]e have an attorney general who likes to sue. It took me a week to get him not to sue Indiana." In the surprised silence that followed this remark, Fred Boissy tried to temporize. He said that "these are problems we've got to talk about, not argue about." Benda replied that Illinois was willing to negotiate but would not be "walked on by Indiana."[16]

Benda's aggressive stance at this meeting was mostly rhetorical. No conclusive study of sedimentation in the Illinois part of the Kankakee had been done. Residents near Momence had watched with alarm for years as sandbars grew in the shallows of the river, especially after the marsh was drained. Citizens believed that sediment from draining the marsh had washed into Illinois, and that bank erosion along the Kankakee was still sending sand across the state line. However, the Task Force report that Benda passed out was equivocal about blaming Indiana for this. In fact, the report acknowledged that "uncertainty remains as to the magnitude and the source of the problem." The Task Force recommended that Illinois begin monitoring sediment at the state line to verify whether or not Indiana was, indeed, to blame.[17]

Press coverage of the March meeting in Valparaiso was extensive. It was also an object lesson in the difficulties the KRBC faced in educating the public about the Kankakee's problems, and about proposals to manage those problems. Six area newspapers published eight articles about the meeting, all with a different focus. *The Times* (Munster, IN) and the *Rensselaer Republican* each contributed two articles. Both the *Rensselaer Republican*'s lengthy articles ignored completely any discussion of the commission's plans for flood control, flood preparedness, or any other substantive river issue, focusing instead on procedural matters. *The Times*, *The LaPorte Herald Argus*, *The Post Tribune* (Gary, IN), the *South Bend Tribune*, and *The Valparaiso Vidette-Messenger* all provided varied accounts about the river management plan proposed by the commission. The press seemed unable to reach a consensus on the specifics of the plan, perhaps because the commission itself was unsure. Only *The Times* and the *Vidette-Messenger* reported the presentation by George Benda, and only *The Times* provided direct quotes that clearly showed Benda's tense exchange with Fred Boissy.

The issue that Benda raised was important, and newspapers should have paid attention. Sand in the Kankakee River was a vexing question. There is little doubt that draining the Kankakee wetlands caused sediment, especially sand, to wash downstream. In 1883, Engineer John Campbell knew that channelizing the river would carry "large quantities of soil and sand into Illinois." He pointed to evidence in the river near Momence that this had already happened after Beaver Lake was drained into the Kankakee.[18] Engineers in 1909 reported huge quantities of sand from the channelized Yellow River accumulating at its mouth, where it joined the Kankakee.[19] It should be remembered that the entire upper river had to be re-dredged after 1915 because of excessive sand buildup in the channel that had been dredged only a few years earlier, sand buildup that had *decreased* the depth of the river by half.[20] And a 1931 report from the US Army Corps of Engineers specifically blamed channelization in Indiana for sending silt and sand into Illinois that "lodged among the trees, and created numerous bars in the river bed."[21]

The questions for planners in the late twentieth century were whether significant sand and soil were still flowing down the Kankakee from Indiana into Illinois, and what effect that sediment might be having on the river in Illinois. During 1978 and 1979, the Illinois Department of Energy and Natural Resources conducted studies that shed some light on these questions. Results were published in 1981 and will be discussed in a later chapter.

Flooding along the Kankakee and the Yellow River was expected to be worse than usual in 1978, and it was. On March 23, the *South Bend Tribune* showed photos of streets in Plymouth that were underwater and reported that some families had left their homes. Rain was expected to increase floods caused by melting snow, and sandbagging had begun in Starke County.[22] A few days later, state officials and others toured the flood area by helicopter and reported that "thousands of acres were underwater," estimating that damages would be in the "millions." The Yellow River continued to rise.[23] Flooding in Lake County was especially severe. Overflow from the Kankakee closed the main road to Shelby before the river crested, and more flooding was expected. By April 3, much of Schneider was underwater and residents were evacuated. More than two thousand acres of cropland flooded, and more rain was expected. Three days later, two hundred homes were affected in the area. An aerial inspection by the US Army Corps of Engineers showed that floodwaters covered ten square miles. Dikes broke between Shelby and Schneider, and broke again after they were repaired. The water level did not begin to drop until late April, a month after flooding had begun. Over ten thousand acres had been inundated.[24]

New flooding inspired familiar complaints. An editorial in the Munster *Times* in early April summarized those complaints, along with criticism of the KRBC:

The Kankakee River often overflows its banks, causing cropland and residential damage to portions of eight northwest Indiana counties. And each spring people say, "Somebody ought to do something about the Kankakee." But so far, talking is about the only thing anyone does. In fact, that's about all anyone has done during 100 years of Kankakee problems.

The Indiana legislature created the Kankakee River Basin Commission in order to change that situation, the editorial continued, but many say that the commission has been slow in reaching those goals. Residents along the Kankakee "want to know when the commission will stop studying and start acting. They study it. And then they study it again. The government has spent thousands of dollars studying that river, and meanwhile, people continue to get flooded out. When are they going to start using what they've found out?"

The editorial defended the commissioners, pointing out that they had encountered "complicating factors," and cited conflict between farmers, who wanted to dredge, and conservationists, who wanted to preserve river habitat. And of course, there was Illinois, which opposed any plan—such as dredging or cleaning the channel—that might cause flooding past the state line. The editorial saw no end in sight and concluded on a resigned note: "The conflict remains unresolved and the saga of the Kankakee continues."[25]

Similar conversations were held elsewhere along the river that month. Jim New, a wildlife biologist with the Indiana DNR, reported a meeting he attended where opposing groups debated the best methods of flood control: levees, or "digging the river deeper [dredging], and blowing out the Illinois shelf [destroying the rock ledge near Momence]." One man suggested an analysis that was unique, and singled out a problem that did not include weather or engineering:

> The problem is not too much rain. The problem is that nobody wants the inconvenience of wet fields for even one week in the spring. So we tile and ditch and pump. All this water is forced into a small channel which was never designed to carry so much water so fast. And when we pump, the river fills and when average rainfalls occur, flooding results. And the whole problem stems from nobody caring what happens to people farther downstream. At this very minute, Schneider and Shelby, Indiana are under water, yet 50 miles upstream, every pump that is operable is dumping more water in the river. Any way you look at it, you will never be able to immediately rid one million acres of a 2-inch rainfall overnight. Not without some common sense and some respect for the people downstream.[26]

The matter could not have been put more succinctly, or more accurately. When the same narrow channel that is used for draining is also used for flood control, only more flooding can result. And it is always worse downriver. Engineer William Whitten had made a similar argument nearly one hundred years before: even *average* rainfall can cause a narrowed channel to overflow. The unnamed citizen—from Lake County, where flooding was severe—suggested something quite novel, however. Something that no one had mentioned in over a hundred years of reported public discourse in Indiana about draining and flooding on the marsh: respect for people downstream. Few seemed to recognize that the "saga of the Kankakee" was in large part the long, sad tale of that lack of respect. The man from Lake County had singled out exactly the problem commented on in chapter 5: the inescapable contradiction at the heart of draining land through a river channel. That is, when upstream owners claim the sole right to control river flow, they effectively deny that right to those downstream.

The "complicating factors" cited by the *Times* editorial were brought into sharp focus at a meeting in Jasper County in May, when more than fifty people showed up to hear Fred Boissy, the county's representative on the KRBC, explain the county's role in the commission, which many opposed. The meeting demonstrated that Kankakee farmers, including KRBC commissioners themselves, were unhappy with any plan for flood control that included the construction of parks along the river. A resident opposed to the KRBC said that he had a petition with signatures of over 250 county residents who also objected to the formation of parks. Boissy defended Jasper County's participation on the commission, but agreed with residents about the parks: "We want the same thing," he said. "I would hope not to see this farmland and homes along the Kankakee River turned into one gigantic park. And this is what I am trying to prevent."[27]

The KRBC, however, did not really have a plan for parks, or really for anything else. This led to very unproductive discussions with Illinois officials. At a Bi-State meeting that month with Indiana state agencies and representatives from Illinois, the KRBC had no formal, detailed plan to present. In response to questions from Illinois delegates, John McNamara, member of the commission, had to admit that Indiana could not provide details for any projects. He said, "I think the problem is you are asking specific questions and we don't have any idea what we're doing." Illinois representatives noted Indiana's lack of sediment and water quality records, and the KRBC's "loose adherence" to the suggested plan of the Kankakee Basin Report. A. D. Luers, chair of the KRBC, conceded that the plan is "not absolutely current." That comment brought a sharp reply from Gordon Graves, an Illinois delegate, who scoffed at "a plan that is not a plan. What is going on here?" he demanded. William Staehle, chair of the Bi-State Commission replied, "I thought that was what we were trying to do is figure out what's going on." The meeting ended with an agreement by all parties to exchange information, apparently in the hopes that the next meeting might be more productive.[28]

By summer, many Indiana newspapers were carrying stories about the conflict between Indiana and Illinois over the river. More accurately, at least five different papers carried an Associated Press story about the issue that featured quotes by Indiana and Illinois officials and focused on Illinois's concerns about sand buildup near Momence. The stark difference between the two states' views of the river was underscored by A. D. Luers, chair of the KRBC. According to Luers, "the river was built for draining," and in his view, lack of maintenance was the chief cause of flooding.[29] Luers no doubt meant that the river in Indiana had been channelized, that is, "built" to drain the marsh, but his choice of words was revealing. For many in Indiana, the Kankakee's only reason for existence was to drain farmland—it had no real identity as a river at all. Somehow, the drain was also supposed to act as flood control. When it failed at this, cleaning out vegetation, or dredging, or both, were seen as necessary to carry off more water. Luers suggested that soil conservation practices in Indiana would mitigate sedimentation in the Illinois part of the Kankakee, but the KRBC did not seem to have a plan for soil conservation.

The KRBC knew it was under pressure to formulate a plan—any plan that would show Kankakee residents that it was serious about flood control. In September, the KRBC's executive committee announced just such a plan—one many farmers had advocated. It was simple, straightforward, and focused, unlike the suggested plan of the Kankakee Basin report, which had called for wetlands preservation and the creation of recreation areas in addition to flood control. The executive committee narrowed its recommendation to one project: clearing and snagging the river. This meant removing logs and clearing overhanging branches, a project that Illinois was on record as opposing. A vote on the proposal would come before the whole commission at its October meeting, including a request to the legislature for $750,000 for the project.[30] Illinois replied quickly.

A day after the KRBC's announcement, Frank Beal, acting director of the Illinois Institute of Natural Resources, released a letter to the press accusing Indiana of bad faith. He said that the action by the KRBC was "clearly disrespectful of the good faith effort the two states have made in developing the Interstate River Coordinating Committee." The two states had pledged "to move together in the Kankakee basin," he said, and charged that the action by Indiana was "plainly unilateral." Beal continued:

> Second, and more important, Indiana's efforts to begin snagging and clearing threaten the well-being of the Kankakee River and the citizens of its basin in Illinois. The proposed program of snagging and clearing, as far as we are presently able to judge, will cause increased flooding and sedimentation in the Illinois portion of the river.

He concluded with a plea for cooperation, asking for the development of "a mutually acceptable management program."[31]

The issue came to a head at an acrimonious meeting at the KRBC's offices in Highland a few days later, when about forty officials from both states traded accusations. A. D. Luers, chair of the KRBC, read a prepared letter at the meeting that took exception to Beal's accusation that Indiana's proposal to clear logs from the river was "unilateral." Luers asked whether

> Illinois' proposals to classify 6,500 acres along the main stream in Illinois for wetlands preservation and monitoring sediment, water quality, and water quantity were not also "unilateral." No one in Illinois asked people in Indiana for their comments or approval, nor, as far as I know, did anyone in Indiana send a letter criticizing the "unilateral action" taken.[32]

Luers went on to defend Indiana's plan:

> There are thousands of trees and clumps of tree trunks now lying in the main channels of the Kankakee and Yellow Rivers. The executive committee of the KRBC feels that these should be removed as a first step in the maintenance program of the river ... We will take every step necessary to get action underway to stop the deterioration of our section of the Kankakee River Basin. I live in hopes that someday Illinois officials will see the light, start their own maintenance program, and quit blaming Indiana for all the problems.

Luers promised that Indiana would "use every method we can to 'park' every particle of Indiana soil where it now lies, and retain the water where it now falls."[33]

Frank Beal read his own prepared statement, calling the clearing and snagging proposal a "major channel modification," which, if undertaken without a "full environmental assessment is alarming." Beal charged that the plan "attacks the symptoms of flooding but ignores the problem at its source," and added that implementation of the plan "would destroy the spirit of interstate cooperation." Beal allowed, however, that Luers's statement about "parking every particle of soil" in Indiana could be a starting point for agreement.[34]

The most trenchant comments came from Gordon Graves, an environmentalist who had been a leader in Illinois Kankakee River preservation activities for a number of years. Graves asked the KRBC point-blank to withdraw its proposal. In addition, he demanded that Illinois officials contact their governor to ask Indiana Governor Bowen to "veto any and all funding for work on the Kankakee River in Indiana." Graves went on to lecture Indiana officials about draining the marsh:

> Your problems began 14,000 years ago. Mother Nature provided a vast marsh which you're trying to farm. In 1893 you removed 30 inches of the rock ledge at Momence.

That didn't solve your problems so you dredged the river. That's when our problems began ... Now you want to reclaim more wetlands ... You have not observed soil conservation practices and you're in trouble and want us to pull your chestnuts out of the fire.

Graves threatened to take unilateral action of his own. He said that a movement was being started in Illinois to form a Kankakee River citizens' council. He warned that the Kankakee River Basin Commission was "playing Russian roulette with its constituency," and that people interested in forming such a council might consider proposing

> installation of a dam between Momence and the state line. It would be a sediment trap and could control the high and low flows. Farmers would have to trade in their tractors for boats and their corn for fish food. This is a growing movement presently. We don't want this to happen, but think about it.[35]

The meeting concluded after two hours with an agreement to form a joint "round table" to discuss the tree removal proposal. Presumably, Illinois and Indiana officials would discuss recommendations from this group at the next meeting of the KRBC in October.

At that meeting, however, the commission drew a line in the sand. Sixteen members (out of twenty-four) attended, and the vote to request funding for snagging and clearing was unanimous, with little discussion. If the roundtable formed at the previous meeting had ever met, its deliberations had evidently deadlocked. A. D. Luers said that Indiana was willing to compromise, but that Illinois had not given "one inch." He did not specify *how* Indiana might compromise. N. A. Spann, Lake County Commissioner, added that "we are not defying Illinois. We are convinced this won't hurt Illinois. We just can't keep discussing different alternatives. It's time to get off the stick and act."[36]

Divisions between Illinois and Indiana were deep and intractable. Indiana's proposal to remove logs from the river became the public issue that focused attention away from the profound lack of trust that underlay attitudes of officials from both states. To be sure, Indiana's and Illinois's different ways of looking at the river reflected different attitudes toward nature, but these attitudes were not *inevitably and necessarily* incompatible. Confrontational language from officials representing both states suggests a dynamic rooted in historical mistrust. Illinois officials had not hindered Indiana's project to destroy the rock ledge in 1893. In 1978, they seemed determined not to repeat that mistake.

Many on the KRBC no doubt agreed with chairman A. D. Luers, who told an interviewer from the *Kankakee Valley Post-News*, apparently in reference to Illinois, "I really think man can improve his environment. Some people think you can't." He added:

"We have a great natural resource in the land and if you think of the economy . . . the land ought to remain farmland . . . The commission believes that every effort must be made to keep the river from becoming a meandering stream and to improve the land."[37]

The notion that the only value to the old marsh was economic aligned perfectly with the ideas of nineteenth-century landowners, who had channelized the Kankakee believing with certainty that draining "improved" land that the marsh had "injured." By 1978, many Americans no longer held unquestioning belief in those values. Aside from that, Luers's remarks seem disingenuous. That Indiana's Kankakee lands should "remain farmland" was hardly in dispute, at least not from Illinois. And by 1978, the straightened river had become quite stable, and there was little danger that it would begin to meander. Luers must have known that. His remarks suggest that he seemed to be worried that environmental ideas, fomented in Illinois, somehow threatened farmland in northern Indiana. He acknowledged that drainage interests would have to do "something to satisfy the fish and wildlife people," but he likely agreed with Fred Boissy, a fellow KRBC commissioner, who had told a meeting in Jasper County that "I would hope not to see this farmland and homes along the Kankakee River turned into one gigantic park."[38] Illinois had proposed no such idea, and neither had the authors of the Kankakee River Basin Report. By October of 1978, the KRBC had not identified any plan that would do much to "satisfy the fish and wildlife people."

Many in Illinois would have agreed with Gordon Graves that Indiana's problems with flooding stemmed from a giant historical blunder. But one does not have to disagree with that idea to see that throwing it into the face of Indiana policymakers was not a productive way to solve contemporary problems. Indiana's glib dismissal of Illinois's concerns, however, did nothing to allay fears that Indiana's actions would cause flooding and more sand deposited along the river near Momence. A. D. Luers's promise that Indiana would "use every method we can to 'park' every particle of Indiana soil where it now lies, and retain the water where it now falls" had received cautious support from Illinois's Frank Beal at the September meeting, but the KRBC did not offer any real plan to address the problems of erosion and flooding that Illinois feared.[39] Illinois had been on the receiving end of one of the largest reclamation projects in the history of the Midwest. Nothing in the public record suggests that Indiana officials, including members of the KRBC, had gained much respect for the people downstream.

By the end of 1978, a year and a half after the formation of the KRBC, Indiana was no closer to a workable flood control program than it had been before. The commission had apparently abandoned the comprehensive plan suggested by the Kankakee River Basin Report and decided to pursue merely a "maintenance project" on the river. The stage was set for a protracted standoff that would eventually move the two states into court and allow the US Army Corps of Engineers to assume a major role in the future of the Kankakee River.

# NOTES

1. Dorothy Bernhart, "Kankakee Unit Elects Luers," *South Bend Tribune* (South Bend, IN), July 29, 1977. All newspaper articles, except as noted, were accessed online at Newspapers.com or Newspaperarchives.com.
2. Debi Waits and Louie Stout, "How 'Safe' Project Plan Backfired," *Special Supplement to the Michiana Farmers Journal* (South Bend, IN), August 13, 1979.
3. *Indiana Code, Title 18, City and Town Government, 1977*, Indiana Code Historical Statutes, https://iuidigital.contentdm.oclc.org/digital/collection/IC/id/88963/rec/17.
4. See *The Times* (Munster, IN), "2 States Disagree on River's Future," and "'78 River Plan Gets Board OK," March 17, 1978.
5. Gene Policinski, "Kankakee Bill Law," *Journal and Courier* (Lafayette, IN), April 27, 1977.
6. "Report on the Water and Related Land Resources, Kankakee River Basin, Indiana, State of Indiana" (US Department of Agriculture, US Department of the Interior, 1976), V-6–V-9.
7. "Report on the Water and Related Land Resources, Kankakee River Basin, Indiana," IX–56.
8. Lori Olszweski, "Kankakee Debated," *The Times* (Munster, IN), December 16, 1977.
9. "Report on the Water and Related Land Resources, Kankakee River Basin, Indiana," v.
10. Letter from Roy M. West, chair of the Kankakee County Board, to William J. Andrews, Chair of the Kankakee River Basin Study, December 31, 1975. The letter, summary of the study, and other letters are from the archives of the Kankakee River Basin Commission (henceforth KRBC archives), Kankakee River Basin and Yellow River Basin Development Commission, Valparaiso, IN.
11. Letter from West to Andrews.
12. Kevin Leininger, "Kankakee Valley Flood Curing Plans Unveiled," *The Vidette-Messenger* (Valparaiso, IN), January 13, 1978.
13. "Kankakee Plan Meets Mixed Reaction," *LaPorte Herald-Argus* (LaPorte, IN), January 12, 1978.
14. Leininger, "Kankakee Flood Curing Plans Unveiled."
15. "'78 River Plan Gets Board OK," *The Times* (Munster, IN), March 17, 1978.
16. "2 States Disagree Over River's Future," *The Times* (Munster, IN), March 17, 1978.
17. "Review Summary of the Report to the Governor," in *Kankakee River Task Force Report to Illinois Governor*, 1978, 4–5. From the KRBC archives.
18. John L. Campbell, *Report Upon the Improvement of the Kankakee River and the Drainage of the Marsh Lands of Indiana* (Indianapolis, 1883), 15.
19. W. D. Pence and Morton Downey, "A Report Upon the Drainage of Agricultural Lands in the Kankakee Valley, Indiana, US Department of Agriculture Circular 80" (Government Printing Office, 1909), 8.

20. W. V. Judson, Report of the Board of Engineers for Rivers and Harbors, House of Representatives Document 931, 64th Congress, 1st Session, March 20, 1916, 13.
21. "Report from the Chief of Engineers on the Kankakee River, ILL and IND, Covering Navigation and Flood Control, Power Development, and Irrigation," House of Representatives Document 784, February 1931.
22. "Flooding Problems Hit Starke County," *South Bend Tribune*, March 23, 1978.
23. Richard Salomon, "Starke Flood Damage in Millions, Cook Says," *South Bend Tribune*, March 29, 1978.
24. See articles in the *Post-Tribune* (Gary, IN), March 24, 1978; *Chesterton Tribune*, April 6, 1978; *Rensselaer Republican* (Rensselaer, IN), April 6, 1978; *The Times* (Munster, IN), April 14, 1978; *Post-Tribune*, April 14, 1978; and *The Times* (Munster, IN), April 28, 1978.
25. "Kankakee: Talk Flows On and On," *The Times* (Munster, IN), April 3, 1978.
26. Jim New, "Flooding Raises Questions," *LaPorte Herald-Argus*, April 20, 1978.
27. "River Goals Confuse Jasper," *Post-Tribune*, May 10, 1978.
28. "14 Committees, Boards Study Kankakee," *Rensselaer Republican*, May 16, 1978.
29. Robert Lee Zimmer, "Kankakee River Soils Relations Between States," *Journal and Courier* (Lafayette, IN), August 31, 1978. See the same article, same date, in the *South Bend Tribune*, and *The Times-Mail* (Bedford, IN); see also *Indiana Tipton Tribune*, September 9, 1978, and *The Vidette-Messenger* (Valparaiso, IN), September 6, 1978.
30. Richard Salomon, "Steps Taken to Clear Rivers," *South Bend Tribune*, September 15, 1978.
31. Garth Snow, "Illinois to Answer River Plan," *Rensselaer Republican*, September 21, 1978.
32. "Group to Study Effects of Kankakee Clearing Plan," *Rensselaer Republican*, September 21, 1978.
33. Rosalind Jackler, "River-Snagging Idea Spawns New Group," *The Journal* (Kankakee, IL), September 22, 1978.
34. Jackler, "River-Snagging Idea."
35. Jackler, "River-Snagging Idea."
36. Lori Olszeweski, "Indiana Groups OK River Improvements," *The Times* (Munster, IN), October 20, 1978.
37. Garth Snow, "Kankakee Should be Kept for Drainage," *Kankakee Valley Post-News* (Demotte, IN), October 5, 1978.
38. "River Goals Confuse Jasper," *Post-Tribune*, May 10, 1978.
39. Jackler, "River-Snagging Idea."

# 9

# ENVIRONMENTALISM COMES FOR THE KANKAKEE, 1979–1983

## Indiana Faces Changing Times

THE KANKAKEE FLOODED AGAIN, SERIOUSLY, IN 1979. SNOW WAS HEAVY that year, as it had been in 1978, and reporters in the Kankakee Valley saw the KRBC as a prime source for information about flooding. In an interview with the *South Bend Tribune* in February, A. D. Luers worried about the consequences of an early thaw.[1] On March 1, he warned that dikes along the river had not been maintained. An official with the Lake County Civil Defense agreed, identifying a dike near Shelby as "very weak," and pointing out that it had broken often before, including in 1978.[2] Heavy rains and melting snow caused low-lying areas away from the river to flood first. Families in both Kouts and LaCrosse left their homes to escape rising water, and an entire block in LaCrosse was evacuated when the flood forced 300 gallons of gasoline from an underground tank onto the street.[3]

A day or two later, the Kankakee and Yellow rivers both overflowed. Flooding began on the Yellow River in Plymouth, on the eastern side of the basin, and spread downriver into the Kankakee to Schneider and Shelby near the state line, where levees gave way. Residents were warned they would probably have to leave.[4] A fifty-foot portion of one dike was washed out, and Luers warned that drainage ditches in south Lake County would not be able to contain water flowing through a break in another dike, causing thousands of acres to flood.[5] It was a bad flood year in Illinois, also, where over 300 families fled their homes along the Kankakee.[6]

The flood of 1979 was worse than the one of 1978, and it gave special impetus for officials to do something, anything, to show that they were serious about addressing the problem. In February, the KRBC requested $760,000 from the Indiana government for clearing and snagging on the Kankakee. In late March, about the same time that floodwaters receded, the General Assembly authorized $600,000 for flood prevention. Senator Niemeyer, who shepherded the bill through the Senate Finance Committee, said that evacuation of families near the river, and repair of dikes by the Indiana National Guard, helped persuade committee members to recommend the appropriation.[7] The KRBC did not make an official announcement for months about how the funds would be used, but environmental groups were quick to pounce. Opposition began immediately.

The Izaak Walton League had been vocal in its opposition to clearing the river since the idea was first introduced. The League's criticism of the KRBC, in fact, had not abated since early 1977, when the commission was approved by the Indiana legislature. In an early 1979 story in the *South Bend Tribune*, Tom Dustin, spokesman for the League, compared the commission to a dragon: "There it is, about to gobble us up, and most people are unaware of it." Dustin claimed that the "power of the KRBC is enormous. They have been given the authority to coordinate the development of the Kankakee River Basin and all the streams that run into it." Ray Guard, writer of the article, apparently sympathized with Dustin's fears, and he liked the colorful metaphor enough to extend it: "The Kankakee Commission, a monolithic body which has been awake but toothless for about a year since the Indiana General Assembly created it, may suddenly rise and shudder with dangerous life," he wrote. Lee Wolf of the Hoosier Conservation Defense League echoed Dustin's concerns. Wolf believed the commission wanted to cut back timber 75 feet on the banks of the Yellow River, and dredge the river. "Drainage," he said. "That's all they can think of. Get rid of the water."[8]

The Walton League and the Conservation Defense League were joined by the Indiana Conservation Council, the largest conservation organization in the state. The council was not opposed to dike repair, but James Rice, executive secretary, sharply criticized clearing and snagging of the river. Rice said that "such a project would ruin the river without providing relief from flooding." The project would "wreck" the Kankakee as a fishing stream, he claimed (though he did not explain why), and he believed that it would do "very little—perhaps nothing—to alleviate flooding."[9]

Lee Wolf was correct in thinking that the KRBC's primary mission was keeping Indiana farmland dry. And there were those who agreed with James Rice that clearing and snagging would do little to actually lessen flooding—that issue will be discussed later in this chapter. But the idea that the KRBC had "enormous power," or that it was "monolithic," is so strange as to be laughable. At their April meeting, members of the commission could not agree on a plan to distribute the $600,000 among the eight counties they represented. Once they did work out a plan, they could not act on it. Any plan had to be approved by the Indiana Department of Natural Resources and by the legislative budget committee. If the commission did not have a plan ready by the time the budget committee met in July, the money could be lost. In addition, Illinois was demanding that the KRBC allow its Institute of Natural Resources to review any plan to clear and snag the river.[10]

Illinois wanted more than a review of the plan. It wanted the power to share in its design. The "Memorandum of Understanding" offered to the KRBC called for permits to be issued by the Indiana DNR, even though none were required. The Memorandum also stipulated that Illinois officials be involved in planning the project—to monitor the work for possible sand flow—and that agencies from both states sign a legally binding agreement on the work. In other words, Illinois officials wanted effective veto power over the plan if they did not approve it. Indiana agreed that Illinois could review, but insisted on the monitoring of sand flow. At a meeting with the KRBC, Byron Wallace of the Illinois of the Illinois Regional Planning Commission told the commissioners, "[w]e won't buy it." The KRBC knew it was risking a lawsuit by refusing Illinois demands.[11]

By June, the commission had a plan for allocating a portion of the funds to each county in the Kankakee basin. Final approval from the DNR was expected that month. A meeting was scheduled with the State Budget Agency, at which the commissioners had to prove that the money was actually needed—this in spite of the fact that the legislature had already approved the appropriation.[12] But the KRBC also had to deal with county officials who balked at any plan that required them to spend local money. The legislature had passed a measure for continuing flood control that called for counties to raise $300,000 each year through taxes. When A. D. Luers told Lake County officials that $150,000 was available for immediate levee repairs (which were badly needed—levees were in poor shape and had broken two years in a row), they objected. They wanted to know why the funds had to be spent on the levees, rather than other projects. They asked Luers why the Army Corps of Engineers could not do the repair work, or why the KRBC could not request federal funds for flood control so they could get around the state requirement to raise county taxes.[13] It was a familiar story. The state had refused to pay for reclamation work a hundred or more years before, and it did not want to pay for flood control in 1979. Counties had resisted assessments for drainage in the past, and now they resisted raising taxes for projects needed to curb the consequences of the massive environmental depredations they had financed, however reluctantly.

The KRBC seemed to be in an impossible position. Few were happy with the program to clear and snag the river, and even the plan to repair dikes drew criticism. The Indiana General Assembly may have created the commission to help manage the river, but they had given it few resources to accomplish that goal. Unwittingly (probably), legislators *had* created a commission that drew intense public scrutiny away from the failures of state government to address flooding and other problems in the basin. The Kankakee River Basin Commission became a target for environmentalists, Illinois officials, the press in and out of Indiana, and Kankakee residents who had suffered annual floods, sometimes severe, for over half a century without real remedy. In 1979, there may or may not have been hope for workable flood control—but the KRBC had become extremely sensitive to complaints that seemingly little, or nothing, was being done, again. Small wonder that the commission had settled on a clearing and snagging program—unpopular as it was—not only to show that something was, indeed, finally being done, but also that it was *possible* to do something, and that the KRBC was the right organization for that something.

Criticism of the plan, and of the KRBC, mounted through the summer. The Indiana Department of Natural Resources officially endorsed the KRBC's plan, but some members of the department did not. The Fish and Wildlife Division opposed it unequivocally, and several members showed up at a KRBC meeting in protest. Biologist Jim New told the commission: "In Fish and Wildlife's opinion, you're going to go in and seriously damage 80 miles of the Kankakee River." New explained that "the trees and snags should remain ... [T]hey provide cool, shaded spots in the river that help lower the water temperature during hot weather. If trees and snags are removed, the resultant higher water temperature will harm the fish." New added that the project's effects on the Yellow River would be more pronounced than on the Kankakee.[14]

An editorial in the Kankakee (IL) *Daily Journal*, headlined "Don't Shove the River," criticized Indiana's lack of cooperation with Illinois and cited Jim New's concerns:

> Bi-state cooperation seems to have become a casualty of Indiana's unilateral planning; consultation with Illinois was not even mentioned by the Indiana Kankakee River Basin Commission as its members met last Thursday ... Indiana's steam-roller like determination to proceed with its project has overridden widespread protests, including those voiced by the Division of Fisheries and Wildlife, a unit of its own Department of Natural Resources, and conservationists from both states.[15]

In August, an in-depth analysis of the KRBC and its controversial flood relief plan was published in a four-page supplement to the *Michiana Farmers Journal*. Entitled "The Kankakee Project: Boon—or Boondoggle?," the eight articles in the special section traced not only the history of Illinois and environmental opposition to the plan, but the history of the plan's creation, its prospects for success, the role of the US Army Corps of Engineers in the controversy, Indiana farmers' reaction to environmental concerns, and the opposition of members of the Indiana DNR.

The articles—originally published as a series in the Michigan City, IN, *News-Dispatch*—were a testament to the growing importance of the KRBC, and to the editors' sense of the "vital importance" of the issue of flood control, and the activities of the KRBC, to their readers.[16] However, the writers of the series were not optimistic about the commission's chances of success, claiming that it was "pitching furiously in a no-win match against farmers, conservationists, the Indiana legislature, and the state of Illinois."[17]

Predictably, farmers were very critical of environmental concerns, one man commenting that the "theories" (about habitat loss) of the Fish and Wildlife biologists were "stupid." He added: "Flood control means everything. There's got to be something done. If these conservationists want to eat fish, let 'em live like the Indians, if they don't want beef and bread and corn."[18] According to reporter Louie Stout, the project had caused "bitter feuding and embarrassment" within the DNR:

> The project has become so controversial some DNR officials have been ordered to restrict their remarks to professional opinions. As opposition to the project mounted, Fish and Wildlife officials were told to "back off" and work with the department in support of the project.

DNR wildlife officials seemed unmoved by flood control concerns. Biologists Jim New and Bob Robertson believed that fish and wildlife of the Kankakee "faced a severe blow" from the project. New commented, "From a wildlife standpoint, this project is detrimental ... and it seems there has [sic] been no efforts at mitigating losses."[19]

The most interesting and revealing comments, however, were from KRBC commissioners themselves, who were dubious about the project's success. William Tanke, commissioner from Porter County, criticized the apportioning of funds to each county in the basin. Tanke said the money would be better spent in the counties hardest hit by flooding, because the $600,000 allocated was not enough to completely clear both rivers. Tanke said that the KRBC's approach was like "putting a band-aid on an open, bleeding wound." Commission chair A. D. Luers agreed that the project would not help a lot with flood control. He affirmed that the KRBC "had tired of empty rhetoric and calls for more studies and surveys, and wanted quick results." Commissioners felt they needed "an action program," he said.[20]

Other commissioners believed that the clearing and snagging project was not the real point. Fred Boissy said that the prime benefit was that the state had finally seen that "we've got a real problem. The $600,000 won't accomplish much, but it's a start." Commission member John McNamara also thought the money was not enough for substantive flood control, but believed, somewhat illogically, that the legislature would be receptive to funding more projects in the future after seeing results of clearing logs from the river.[21] Whatever their reservations about the effectiveness of clearing and snagging, however, commissioners would unequivocally advocate for it against Illinois and environmentalist objections for the next three years.

The one agency that might hold real power over the life or death of the project had not been heard from when these articles were originally written. An editor's note, included in the special supplement, sounded an ominous note about that agency. The US Army Corps of Engineers was requesting an application for a permit for the clearing and snagging project from the KRBC. According to the editor's note, Illinois had asked the Corps to investigate the project.[22] This event would be far more consequential than anyone expected.

Environmental regulation was hardly the traditional mission of the US Army Corps of Engineers. Established by President Thomas Jefferson in 1802 to ensure the navigability of American waters, its role expanded quickly. By the late twentieth century, the Corps had a long history of constructing large, expensive river projects that were often much more notable for their environmental damage than conservationist protection. In his comprehensive study of American rivers, Daniel McCool claims that the Corps, allied with Congress, had "literally changed the face of riverine America by building 11,750 miles of levees, 12,000 miles of navigation channels, 276 locks, 350 hydropower facilities, 926 harbors, and 692 dams."[23] Many of these projects were controversial. McCool points out that they were "sometimes in the national interest, occasionally in accordance with sound economic principles, but rarely built in an environmentally sensitive manner, and sometimes a gross waste of money."[24]

McCool writes that the Corps began a new life in 1986 when it was forced by Congress to do fish and wildlife mitigation in addition to project construction, and became a "schizophrenic agency, tearing up rivers with one hand, and restoring and preserving rivers with the other." He cites the Kissimmee River project, referred to in chapter 7, as an example of the Corps's new-found mission for restoration.[25] However, the Corps began protecting waterways as early as the 1970s, and the reason for that lay in seismic social and political changes of the 1960s and 1970s.

Conservationists first found their voices in the early years of the twentieth century, as outlined in chapter 6, when government and private groups united to advocate for an awareness of the fragility of the American landscape. The Franklin Roosevelt administration had been especially active in restoration of natural resources through organizations such as the Civilian Conservation Corps. The environmentalism of the 1960s and 70s was different. It was certainly about awareness and restoration, but it was also a social movement whose goal was not only to awaken recognition of environmental loss, but also to force serious changes in national behavior. It was a decades-long effort to reverse the damages of a century or more of environmental abuse, and to ensure a future for environmental protection. Environmentalism was not just a matter of speeches and protest. It arrived in force on the doorstep of the US Congress.

Between 1963 and 1978, Congress passed a dozen or more pieces of major legislation, all either designed to protect wildlife and natural places or to regulate contamination of air, water, and soil. This legislation included the Clean Air Act of 1963, which permanently funded state pollution control agencies; the Wilderness Act of 1964, which protected certain areas as "wilderness"; and the Water Quality Control Act of 1965, which allowed the federal government to establish water quality standards. Two pieces of landmark legislation came along in 1969 and 1970. The National Environmental Policy Act of 1969 required federal agencies to prepare an environmental impact statement for every project or law that would affect the quality of the human environment. In 1970 the Environmental Protection Agency was founded to regulate air and water quality, biohazards, and solid-waste disposal. The Federal Water Pollution Act of 1972 gave pollution control of waterways to the federal government. Other legislation in the 1970s protected endangered species and wilderness, and gave the federal government the power to set grazing, preservation, and mining policies on public land.[26]

Public sentiment gave special impetus to the creation of these laws. On the first Earth Day in 1970, an estimated 20 million Americans attended events at tens of thousands of sites including public schools, universities, and communities. That environmental concern continued. According to the Library of Congress, more than 200 million people worldwide had participated in Earth Day celebrations within twenty years.[27]

Like many other federal agencies in the 1970s, the Army Corps of Engineers was obligated under law to keep a sharp eye on any project that might threaten the natural environment. Not long after endorsing the snag and clear idea, the KRBC had anticipated the possibility of needing a permit from the Corps. In November 1978, the commission contacted the Corps's district office in Chicago through Indiana Senator Richard Lugar's office, asking for information about requirements. The Corps responded that the Corps had jurisdiction over waterways under two laws: Section 10 of the River and Harbor Act, which pertained to navigable streams (only a short length of the Kankakee in Illinois was deemed navigable), and 404 of the Clean Water Act. Section 404 empowered the Corps to issue permits for the "discharge of dredged or fill material into navigable waters, up to their headwaters, including adjacent wetlands." Indiana might need a Section 404 permit for its work. The KRBC would have to inform the Corps of disposal site locations along the river, and the Corps would determine if the locations were in wetlands under Corps jurisdiction.[28] By August 1979, the KRBC had not applied for a permit to clear and snag the Kankakee River. The Corps had decided, however, that it needed a detailed description of the plan, and it warned that an Environmental Impact Statement might also be necessary. If so, the project could be delayed for as much as a year.[29]

Undeterred, in October the KRBC published bids for the clearing and snagging work, and for levee repair. Donald Potter of *The Times* marked the occasion as a historical moment, writing: "For the first time since the Kankakee River was straightened in 1917, state-funded river control work is set to begin by January."[30] Potter was wrong about the year that river channelization ended, but he was right about the complete absence of state-sponsored flood work since the marsh had been drained. KRBC members seemed confident that their plans for dike repair and clearing and snagging could be completed by early 1980. Illinois had other plans, however, and officials there knew how to apply the right kind of pressure. In December, the state sued the Army Corps of Engineers, charging that the Corps was not doing enough to protect the Kankakee River.[31] A few days later, the Corps issued a cease and desist order to stop part of the levee repair in Lake County.[32] Indiana found itself in a fight for control of the Kankakee River with the US Army Corps of Engineers—in effect, with the federal government. It was a fight the state could not win, but fight it did, for years. The issue would not be fully resolved until 1983.

The Corps claimed that Indiana did not have a permit and had illegally filled in wetlands at locations near the levees. The KRBC balked at stopping work—which was almost completed—and produced a letter from the Corps showing that a permit was not needed. Commissioners also charged that the Corps had issued the stop order because of Illinois's suit against them. In a meeting with the KRBC, Colonel Howard Nicholas, the Corps district engineer, denied that Illinois had anything to do with his order, and told the KRBC that if work did not stop, the Corps could demand the removal of the levees. He said that his staff had made a mistake in recommending levee repair work. The KRBC Executive Committee voted to finish the work regardless of the orders, but Nicholas told them that he would file suit with the US Attorney (General) if they disobeyed him.[33]

The Corps certainly did seem confused. After the meeting, a Corps official told A. D. Luers that there was "a complete lack of communication between two factions of the Corps."[34] In August, the Corps had demanded a detailed description of the project and threatened to require an environmental impact statement, then in December, issued a letter saying no permit was needed. Now in January, it was requiring a permit and threatening to sue the KRBC if it did not stop work immediately on the levees. No wonder the commissioners felt whipsawed. Few believed that the Army Corps of Engineers was not working in Illinois's interests. The Corps may have been confused, but it was also serious about regulating the Kankakee: it had begun a navigability study of the river in Indiana. If navigability was confirmed, the Corps would have jurisdiction over the entire Kankakee River. The KRBC could see the handwriting on the wall. A few days after the meeting with Colonel Howard, the commissioners stopped work on the levees and applied for a Section 10 permit to clear and snag the river.[35] By the end of January, the Indiana Attorney General had become involved, agreeing to defend the KRBC to the US Army Corps of Engineers.

The commission was going to need all the help it could get. Other federal agencies were quick to join with the Corps against the KRBC. In February, the US Fish and Wildlife Service and the Environmental Protection Agency agreed with the Corps that the levees endangered wildlife. Both agencies advised the Corps that they wanted to make recommendations on Indiana's permit application.[36] A few months later, events took an even more serious turn.

The dispute over flood control on the Kankakee River could no longer be confined to meetings between Indiana and Illinois officials, or to a disagreement over how federal regulators communicated with KRBC commissioners. Antagonisms that had been simmering for years broke into open legal warfare. In April, the Illinois attorney general added the KRBC to the suit against the Corps, naming sixteen of its members as defendants, and sought a court order to remove levee improvements in Indiana.[37] In May, the Izaak Walton League and Citizens for a Better Environment piled on, seeking to join Illinois in its suit against the KRBC and the Corps. The Waltons claimed that the KRBC had destroyed trees on land near the river owned by the League, and sought extensive compensation from two commissioners. The League also charged that work on levees had stripped away vegetation at the levee construction site, and that fill material was caving into the river, "causing siltation problems and setting the stage for additional dike problems and obstruction of the river's floodway." The League's suit was personal: it demanded fines of $10,000 a day for each commissioner since December.[38]

On the advice of the attorney general's office, the KRBC halted all work to clear logjams from the river because of Illinois's suit against the commission. However, the commissioners decided that a counterattack was the best defense. They requested that the attorney general sue Illinois for causing increased flooding in Indiana, accusing Illinois residents of illegally building bridges and other obstructions that impeded river flow.[39] In the struggle for legal one-upmanship, however, Illinois was way ahead of Indiana. A month later, an injunction officially prohibiting all flood control in Indiana was issued by a federal judge in Chicago.[40] Illinois had won the first round: the KRBC could not resume any work on levees or logjam removal until the injunction was lifted. Indiana did win a victory in September, when the Corps decided the Kankakee was not navigable in Indiana. This decision considerably lessened the power of the Corps to limit flood control, but Illinois immediately challenged the ruling, suing the Corps to declare the entire river navigable.[41]

The KRBC's legal situation did not improve the following year. In January 1981, a US District court settled the navigability question, ruling that all of the Kankakee River was navigable, a win for Illinois.[42] The US Army Corps of Engineers now had undisputed jurisdiction over the entire river, from its headwaters in Indiana to its mouth in Illinois. This meant that the KRBC could not proceed with any flood control work without permission from the Corps. Public hearings on permits for levee repair work and clearing logs from the river began in February.[43]

The first meeting, in Indiana, drew over 200 people, mostly Kankakee farmers but also elected officials from both states, lawyers, scientists, environmentalists, and members of the KRBC. It was a long meeting. For four hours, Kankakee residents made their case for the necessity of flood control while environmentalists and scientists from Illinois, Indiana, and federal regulatory agencies spoke about the damages to wildlife and wetlands that clearing of vegetation on the river would cause. Michael Aylesworth, who had succeeded A. D. Luers as KRBC chair, put the case for flood control in a familiar context, prioritizing the value of farmland over the value of wildlife. When the river was channelized, he said, "the unique wetland marsh was lost forever, but now we must face reality. If remedial work isn't allowed to be done, then valuable farmland will be endangered and threatened at a pace which will be intolerable." The magnitude of the threat to farmland was never really explained, but a Purdue economist was ready with statistics about economic value. He placed the value of farmland in the Kankakee basin at more than $3 billion, and claimed its annual crop yield was worth almost $500 million. The implication was clear: the "reality" of an economic enterprise worth nearly half a billion dollars was somehow seriously threatened (alas, "endangered"!) by a desire to mitigate damage to wildlife on the Kankakee River in Indiana and Illinois.[44]

At the second meeting, in Kankakee, Illinois, the state produced experts who testified to the environmental consequences of allowing the clearing and snagging project to proceed. Geologists showed that the channel in Indiana was stable—there was little danger that the river would begin to re-meander. Although a great deal of sand had washed into Illinois from Indiana in the first half of the twentieth century, the sand load had decreased since. However, clearing and snagging in Indiana would damage river wildlife in Illinois because an increase in sand and silt would be "inevitable." A hydraulic engineer with the Illinois Board of Survey concurred, predicting that clearing of vegetation would increase erosion along the banks of the river in Indiana. A biologist pointed out that aquatic life that depended on the rocky river bottom in Illinois would not survive the increased sand produced by clearing the river in Indiana.[45]

Some in Illinois worried about flooding, but the major division between the two sides could not have been clearer. Illinois opposed any project that harmed wildlife, and was willing to commit considerable state resources to stop such a project. Kankakee landowners and officials in Indiana, including the KRBC, had but one priority: the protection of farmland, even when they could not make entirely clear exactly how the "maintenance" project they planned would actually achieve their goal or, indeed, how serious the threat really was. No one had spoken for the myriads of wildlife or thousands of acres of habitat that were destroyed by draining the marsh between 1901 and 1923. Officials and landowners in Indiana would have been completely nonplussed if they had. In 1981, they still were.

In April, the Corps increased pressure on the KRBC by requiring additional information for permits. The Corps wanted all trees intended for removal to be labeled on aerial photographs, and it wanted a fishery inventory—a very extensive fishery inventory. According to the Corps, the study should show "the relationships [of fish species] between the main channel and old channel, backwater and tributary area; [and] quantify utilization of snags (fallen trees, log jams, and debris) in the river for [fish] food sources." The instructions for tree photographs were also detailed. The Corps required a "narrative accompanying the photographs detailing site by site, the number of live trees and the number of dead trees which will be cleared, the number of individual snags, the number of trees in log jams in the river and the approximate amount of additional debris to be removed from each site." Additional information requested included the method for removing trees and an explanation of how live trees were to be protected in the removal process.[46]

KRBC commissioners knew harassment when they saw it. They complained that the Corps's new demands were "ludicrous," but they voted to provide the information.[47] For months they had believed that the Corps's Chicago office favored Illinois in its efforts to block flood control in Indiana. KRBC chair Michael Aylesworth appealed to Indiana senators Richard Lugar and Dan Quayle to have Corps jurisdiction of the Kankakee transferred to the Detroit district.[48] In May, the transfer was approved but the Chicago office's permit requirements had to be met. That month, the KRBC began marking trees on a hundred-mile stretch of the Kankakee.[49]

Commissioners may have thought that Corps's demands were unfair, but they made a good-faith effort to comply with those demands. Marking of trees took weeks, extending well into June. Connie Weir, project director of the KRBC, along with five DNR employees, navigated the Kankakee in two boats, one for each riverbank, marking trees with paint guns. When they were finished, they had marked and mapped 30,000 trees for removal—living and dead.[50] Commissioners may have been ambivalent (at best) about the effectiveness of clearing trees in preventing floods, but they were serious about getting permission to do it.

Flooding on the Kankakee and elsewhere in Indiana was extensive that year. Somehow, Connie Weir and her crew managed to work in spite of conditions that were almost as bad as 1979. Rainfall in some parts of Indiana was ten times the normal amount in June, averaging five inches for a week in the northern part of the state. Over a hundred people were evacuated from Shelby, but the dike there held.[51] Levees elsewhere failed. Some gave way in Jasper County, allowing one thousand acres to flood. The Little Calumet River in Lake County also flooded, and Lake County, once more, was designated a disaster area.[52] Losses to cropland were heavy. According to the *South Bend Tribune*, damages totaled between $125 and $150 million.[53]

Summer brought good news for the KRBC, however. In July a federal judge in Chicago dismissed Illinois's $150 million suit against Indiana, but with a crucial caveat: Indiana must have permits from the US Army Corps of Engineers to proceed with clearing the river and repairing levees.[54] August brought more good news when the Corps granted the KRBC a permit for levee repairs. Apparently, Aylesworth's work with Indiana senators paid off, because the announcement came from the offices of Senators Lugar and Quayle. A Lugar aide declared confidently, but prematurely, that the granting of the permit established that Indiana had the right to determine what kind of maintenance was appropriate for the state's portion of the Kankakee River.[55] The aide apparently forgot that the US Army Corps of Engineers had not yet ruled on the question of whether Indiana could clear trees and logjams from the river. Indiana's unilateral right to manage the Kankakee was still very much in play. By the end of 1981, the Corps still had not made that ruling.

In the struggle for control of the Kankakee, Indiana may have won an important battle when Illinois's suit against the state was dismissed, but the war of words was far from over. Emotions still ran high—very high. In March 1982, a public meeting was held in Kankakee, Illinois, to discuss the results of a years-long study of sedimentation in the Illinois portion of the Kankakee. Begun in 1978, the study was an outgrowth of the Kankakee River Task Force's report to the Illinois governor that focused on sand deposits in the Kankakee west of the state line, discussed in chapter 8. Scientists who spoke at this meeting were the same who had spoken at the Army Corps of Engineers hearing almost exactly a year before, and they presented some of the same information, but this meeting was not cordial. Illinois residents believed that Indiana was the primary source of sedimentation on their side of the state line, and they wanted a scientific report that proved it.

According to the *Rensselaer Republican*, the crowd was openly hostile to officials from Indiana who attended the meeting. One man shouted at Larry Wickert, chair of Starke County Commissioners, "do you want to get out of town alive?" Wickert replied that there were people in Indiana who wanted to "go to war" also, but that Indiana was interested in erosion control, and wanted to reduce sedimentation. A man from Momence challenged the Indiana residents' right to be at the meeting, asking if they had been invited. Attorney Ralph Bower waved a letter of invitation from an Illinois agency and said, "if anyone wants to throw me out after I've been invited, they're welcome to me."[56]

The crowd was almost as hostile to the study—prepared by Illinois engineers and scientists—as it was to representatives from Indiana. Published as *The Kankakee Yesterday and Today*, the report focused on the geology of the river, sedimentation, and the effects of sand on aquatic life. Its conclusions pleased no one. The audience wanted an "action plan"—a report that explained how to reduce sedimentation and flooding from Indiana. They wanted recommendations for erosion control east of the state line.

They were disappointed. Nani Bhowmik, river hydrologist with the Illinois State Water Survey, showed that some of the sand in the river originated in Illinois, not Indiana, and the Iroquois River (tributary to the Kankakee in Illinois) actually carried much more sediment than the Kankakee. Allison Brigham, of the Department of Energy and Natural Resources, had better news for those who wanted to stop Indiana from clearing and snagging the river. Brigham "predicted a reduction in diversity and abundance of aquatic life" if clearing of logs and other vegetation was permitted. An increase in sedimentation caused by that activity could kill off 30 percent of resident aquatic species, and 70 percent of the total the number of fish, she warned.[57]

Most of the audience remained convinced that Indiana farming practices caused sedimentation in the river. They weren't wrong; the report showed evidence of significant sand deposits between 1939 and 1954.[58] (The report leaves unstated the implication that these deposits could have been caused by the draining of the marsh.) However, sedimentation had decreased quite a bit after 1954, indicating that the river was in a state of equilibrium. In fact, the banks of the river in Indiana were relatively free of erosion.[59] Gordon Graves, environmental nemesis of the KRBC, pointed out that the study had ignored citizen contributions. His remarks drew a round of applause.[60]

Reports from scientists and engineers did little to change minds on either side of the state line. Farmers in Indiana were convinced that biologists were "stupid" for wanting to curtail flood control projects that could damage wildlife. Residents along the river in Illinois saw themselves as victims in an eighty-year-old struggle with Indiana landowners who refused to take responsibility for problems their actions caused downriver. Both camps were aggrieved. Both were entrenched.

The acrimony of the meeting in Kankakee obscured the Illinois scientists' analysis of Indiana's snagging and clearing proposal. In a thoughtful series of articles about the study, Robert Themer, environmental reporter for the Kankakee, Illinois, *Daily Journal*, put those conclusions in the context of that proposal. According to Themer, geologist David Gross and hydrologist Nani Bhomik stated the matter clearly. "The dominant thought in our minds was in reaction to clearing and snagging upstream," Gross said. Themer summarized the scientific reasoning for opposing Indiana's project:

> Indiana drainage advocates claim their proposed work will decrease the flow of damaging sand into Illinois. The snags and overhanging trees cause erosion of the opposite banks, they say ... but the Hoosiers are wrong. "The river banks in Indiana are almost exclusively sand," Gross said. "... [T]he river system in Indiana runs through sand, sand, and sand." And, while it is unusual for a channelized river to become stable, that is what has occurred on the Kankakee in Indiana ... The stability developed because the channel banks were allowed to grow up in trees and other vegetation, Bhowmik said. "Where there are trees, the banks are extremely stable. Where there are not trees, it is unstable," he said. [Bhowmik added that] the Indiana proposal to clear trees and snags at some 1,700 locations will destabilize large areas of the banks. [He estimated] that the work, if allowed, would increase the amount of river-borne sand coming into Illinois by 30–40 percent. It would also result in the destruction of levees which provide flood control in Indiana.[61]

Themer's point is that when carefully considered, the study so vilified by Illinois residents was actually an extended geological, hydrological, and biological refutation of the KRBC's claims for the necessity, and the benefits of clearing and snagging the Kankakee River in Indiana.

As it turned out, the US Army Corps of Engineers agreed with the Illinois scientists. By September, the Corps's Final Environmental Statement was complete. The report concluded that

> [the] proposed project would cause significant impacts on 1) vegetation, erosion, levee stability and safety, 2) fisheries of the river, 3) two federally listed endangered species, the Indiana bat and bald eagle, and 4) several other wildlife species, especially cavity nesters. The proposed clearing and snagging would also result in minor adverse impacts on 1) water quality, 2) agriculture and economics, 3) aesthetics, and 4) recreation.

In addition to ecological damage, the Corps specified severe hydrological problems with the proposal, pointing out that "the project would result in negligible flood control benefits." The recommendation for the project was negative because "[t]he proposed clearing and snagging project does not appear to be in the best interests of the public."[62] In December, it was official. Lieutenant Colonel Christos Davos, district engineer of the Chicago office, sent a letter to the KRBC denying the request to clear and snag the river.[63] The KRBC appealed the decision to the division engineer, General Scott Smith, who upheld Davos's decision the following March.[64]

In April 1983, the KRBC decided it was through with the Corps, at least for the time being, and also called it quits on big river projects. In consultation with Indiana legislators and the DNR, the commissioners decided against seeking another permit for a general flood control project on the Kankakee River.[65] The KRBC had first proposed clearing and snagging in 1979—it was the only basin-wide flood control measure on the Kankakee River initiated by any Indiana organization. After four contentious years, the project was dead. How had it failed, and why?

The main reason, of course, was the steadfast opposition from Illinois. The KRBC's first proposal, based on the US Soil Conservation study of 1976—ambitious and comprehensive, but weak and untenable—had almost no traction in Indiana, and Illinois officials mobilized against it. The clearing and snagging project was born of desperation. The KRBC saw it as a workable compromise, a "plan of action" that had the support of Indiana farmers and the General Assembly. Commissioners seriously doubted the plan was an effective measure for flood control, but publicly promoted it as a "maintenance" program to mollify Illinois. Illinois didn't buy it, and neither did environmental groups, in and out of Indiana. Illinois knew how to work the levers of power in the courts, and used that knowledge with the US Army Corps of Engineers. The KRBC was forced onto the defensive in a protracted legal and regulatory fight that it could not win.

Michael Aylesworth, former chair of the KRBC, blamed KRBC's failure on politics. The Corps's rejection of the permit was "a political decision rather than a sound engineering decision," he declared to a Valparaiso *Vidette-Messenger* reporter. The KRBC had "been sandbagged all along" by the Chicago district of the Corps, he said, adding, "[t]hose who live in the basin don't have a say in it. I find it ridiculous." Without the clearing project, he complained, Indiana residents are "left to wallow in our own water."⁶⁶

Aylesworth was right. Politics, of course, played an important part in the struggle for control of the Kankakee River in Indiana after 1977. The political forces in Illinois that halted the clearing and snagging plan used the courts to block the plan, just as Indiana landowners in 1901–23 had used sympathetic county courts to overcome opposition to channelizing the river and draining the marsh. More to the point, politics was an inescapable fact of life for the commissioners. The KRBC was a political organization, after all. All members of the commission were elected officials in their counties, and each was appointed by the governor. The commissioners had been forced to negotiate a formidable political terrain from the beginning—a terrain that included more obstacles than Illinois environmentalists. Indiana farmers had objected to the first proposal for flood control offered by the commission, and the legislature itself—no doubt for political reasons—weakened the commission by underfunding it.

Yes, politics was to blame for the KRBC's defeat. But commissioners were opposed to the idea of parks bordering the Kankakee, and they were consistently dismissive of the concerns of those they referred to as the "fish and wildlife people." The KRBC would have to learn how to deal with opposition forces that had at their disposal a huge regulatory apparatus in the service of a new national dispensation: the mitigation of environmental damage. The concept of such damage hardly existed in 1861–1923 when Beaver Lake and the Kankakee Marsh were destroyed. *Mitigating* such damage was a strange concept to many half a century later. After 1983, efforts to control flooding on the damaged Kankakee River would have to take this brave new world into account.

# NOTES

1. Jeff Kurowski, "Official Sees Threat of Floods Near Rivers," *South Bend Tribune* (South Bend, IN), February 19, 1979. All newspaper articles, except as noted, were accessed online at Newspapers.com or Newspaperarchives.com.
2. Donald Potter, "Bad Dikes Raise River Flood Peril," *The Times* (Munster, IN), March 1, 1979.

3. Kevin Leininger, "Flooding Routs Families in South County," *The Vidette-Messenger* (Valparaiso, IN), March 5, 1979.
4. "Floods Threaten 2 Communities," *The Times* (Munster, IN), March 5, 1979.
5. "Residents Flee Flood," *The Times* (Munster, IN), March 6, 1979.
6. "Flood-Weary Midwesterners Get Breather," *Kokomo Tribune* (Kokomo, IN), March 12, 1979.
7. Donald Potter, "Flood Aid Cash Okd," *The Times* (Munster, IN), March 27, 1979.
8. Ray Guard, "Walton League Fighting a 'Monster,'" *South Bend Tribune*, February 11, 1979.
9. "Threat to Kankakee Seen," *The Indianapolis Star*, March 25, 1979.
10. "Kankakee Flood Fund Use Delayed," *The Times* (Munster, IN), April 20, 1979.
11. "Flood Funding Fight Possible," *The Times* (Munster, IN), May 23, 1979.
12. Richard Salomon, "OK Needed on Funds," *South Bend Tribune*, June 3, 1979.
13. "Drain Board Gets Levee Repair Cash," *The Times* (Munster, IN), June 5, 1979.
14. Mark Johnson, "Biologists Protest River Snagging," *LaPorte Herald-Argus* (LaPorte, IN), June 22, 1979.
15. "Don't Shove the River," reprinted as a guest editorial in the *LaPorte Herald-Argus*, from the Kankakee, IL, *Daily Journal*, July 9, 1979.
16. "A Word to Our Readers," *The Michiana Farmers Journal* (South Bend, IN), August 13.
17. Debi Waits and Louie Stout, "How 'Safe' Plan Backfired," *Special Supplement to the Michiana Farmers Journal* (South Bend, IN), August 13, 1979.
18. Waits, "Farmers Criticize Biologists, Stress Drainage Concerns," *Special Supplement to the Michiana Farmers Journal*, August 13, 1979.
19. Stout, "Project Causes Feuding in DNR," *Special Supplement to the Michiana Farmers Journal*, August 13, 1979.
20. Waits, "The KRBC Plan: How It Started, Who Supports It, Who Doesn't," *Special Supplement to the Michiana Farmers Journal*, August 13, 1979.
21. Waits, "The KRBC Plan."
22. "Permit Required," *Special Supplement to the Michiana Farmers Journal*, August 13, 1979. From Kankakee River Basin Commission Archives (henceforth KRBC archives), Kankakee River Basin and Yellow River Basin Development Commission, Valparaiso, IN.
23. Daniel McCool, *River Republic: The Fall and Rise of America's Rivers* (Columbia University Press, 2012), 28.
24. McCool, *River Republic*, 28.
25. McCool, *River Republic*, 31.
26. Carolyn Merchant, *The Columbia Guide to American Environmental Policy* (Columbia University Press, 2002), 261–65.
27. Library of Congress Digital Collections, "Today in History—April 22: Earth Day,"

accessed July 1, 1924, htttps://www.loc.gov/item/today-in-history/april-22/.
28. Letter from James R. C. Miller, District Engineer, US Army Corps of Engineers, to Senator Richard Lugar, November 21, 1978. From KRBC Archives.
29. "Permit Required," *Special Supplement to the Michiana Famer's Journal*, August 13, 1979.
30. Donald Potter, "Kankakee Work Plan Under Study," *The Times* (Munster, IN), October 15, 1979, and "Bids Advertised on Levee Repairs," *The Times* (Munster, IN), October 24, 1979.
31. Mitchell Locin, "Army Corps Sued Over River Protection," *Chicago Tribune*, December 27, 1979.
32. Donald Potter, "Kankakee River Dike Project Work Halted," *The Times* (Munster, IN), December 30, 1979.
33. "Kankakee River Work Continues Despite Orders," *The Times* (Munster, IN), January 3, 1980, and Donald Potter, "Corps May Demand Removal of Levees," *The Times* (Munster, IN), January 9, 1980.
34. Donald Potter, "Corps May Demand Removal of Levees."
35. Donald Potter, "Kankakee Project Halted," *The Times* (Munster, IN), January 11, 1980.
36. "Levees Endanger Wildlife Area," *The Times* (Munster, IN), February 18, 1980.
37. "Scott Adds to Lawsuit," *The Times* (Munster, IN), April 15, 1980, and Donald Potter, "Scott Seeks Court Order," *The Times* (Munster, IN), April 22, 1980.
38. "Indiana IWL Sues Kankakee River Agency, Also Names Corps in Federal Court Action," *Hoosier Waltonian*, Spring 1980. From the KRBC Archive.
39. "River Project Halted," *The Times* (Munster, IN), May 22, 1980.
40. "River Control Halted," *The Times* (Munster, IN), June 15, 1980.
41. Donald Potter, "River Status to be Fought," *The Times* (Munster, IN), September 4, 1980.
42. "River Ruled Navigable," *The Times* (Munster, IN), January 20, 1981.
43. Richard Salomon, "Basin Work in Hearings," *South Bend Tribune*, January 30, 1981.
44. Jim Procter, "War Between States Over Kankakee River Resumes," *Post-Tribune* (Gary, IN), February 25, 1981.
45. Allen Essex, "Illinois vs. Indiana; Kankakee River Subject of Hearings," *Kankakee Valley Post News* (Demotte, IN), March 3, 1981.
46. Richard Salomon, "Corps' Demands 'Ridiculous,'" *South Bend Tribune*, April 17, 1981.
47. Salomon, "Corps' Demands 'Ridiculous,'" *South Bend Tribune*.
48. "Favorable Kankakee Developments Cited," *Post-Tribune* (Gary, IN), May 22, 1981.
49. "Tree Tagging Along River Gets Started," *The Times* (Munster, IN), May 18, 1981.
50. David McCarty, "Controversy Rages on Scenic Kankakee River," *The Indianapolis News*, November 19, 1982.
51. "Bad Weather Continues to Plague Indiana," and "Rainfall 10 Times Normal Amount,"

*Noblesville Ledger* (Noblesville, IN), June 16, 1981.

52. Bob Ashley, "Lake County Designated as Disaster Area," *Post-Tribune* (Gary, IN), June 17, 1981.
53. "Flood Loss $125–$150 Million," *South Bend Tribune*, June 19, 1981.
54. "Judge Dismisses Suit on Kankakee River," *The Post-Tribune* (Gary, IN), July 21, 1981.
55. Lisa Mahoney, "Indiana Wins Right for Kankakee Levee Repairs," *The Times* (Munster, IN), August 2, 1981.
56. Garth Snow, "Kankakee Study Disappoints Illinoisans," *Rensselaer Republican* (Rensselaer, IN), March 25, 1982. From KRBC Archives.
57. Snow, "Kankakee Study Disappoints Illinoisans."
58. J. Loreena Ivens et al., *The Kankakee River Yesterday and Today* (Illinois Department of Natural Resources, 1981), 10.
59. Ivens et al., *The Kankakee River Yesterday and Today*.
60. Snow, "Kankakee Study Disappoints Illinoisans."
61. Robert Themer, "River System Stable, but Fragile," *The Journal* (Kankakee, IL), March 28, 1982. From KRBC archive.
62. Qtd. at length in Jack Perry, "Ikes Commend Corps on Kankakee Ruling," *Post-Tribune* (Gary, IN), September 19, 1982.
63. Diane Donovan, "River Commission Disappointed," *The Times* (Munster, IN), December 5, 1982.
64. James Wensits et al., "Kankakee Project Denied Corps Approval," *South Bend Tribune*, March 29, 1983.
65. Richard Salomon, "Panel to Give Up Kankakee Project," *South Bend Tribune*, April 21, 1983.
66. "Corps Scraps River Clearing," *The Vidette-Messenger*, March 30, 1983.

## 10

# WIDE LEVEES AND THE STRUGGLE FOR RESTORATION, 1983–2002

The Master Plan and the Wildlife Refuge

IN APRIL 1983, THE KRBC MAY HAVE BEEN THROUGH WITH THE ARMY CORPS of Engineers, but the Corps was not through with the KRBC, or with the Kankakee River. And at least one environmental organization still had its eye on the basin, with a keen interest in planning for change. That month, the National Wildlife Federation had an idea for reviving an old study of the river, and it wanted help from the Corps. Ed Osann, director of the federation's water resources program, asked Congress to authorize $200,000 for the Corps to develop a "multi-use plan" for the Kankakee River basin. Osann argued—plausibly—that what was needed was a "comprehensive look at all the resource needs in the basin. Wildlife management, recreation access, groundwater management and erosion control as well as flood control." Osann was thinking about the 1976 study of the basin created by the Indiana DNR and the Soil Conservation Service—the same study that the KRBC had used as a guide for its first proposal for flood control and other measures in 1978. (This plan was rejected by Illinois officials and Indiana farmers, as discussed in chapter 8.) Osann believed that the 1976 study offered a "good point of departure for the preparation of a new feasibility report by the Corps."[1]

The Corps agreed. The KRBC, however, was not in a forgiving mood. At their April meeting, the commissioners rejected the idea. "We don't need any more studies," they said, apparently to "purposely exclude" the Corps because of "frustration over the permit denial." Smarting from its defeat, the commission wanted to focus only on smaller projects.[2] However, the following July, the KRBC listened to a proposal from Colonel Christos Dovas, head of the Corps's Chicago district. Dovas, who had recommended against granting a permit for the clearing and snagging project, defended his rejection of the permit and explained the proposed study.[3]

Dovas pointed out that the KRBC's plan for clearing the river was not "balanced with environmental concerns." Dovas saw himself as "a friend of the river," who was "vitally concerned with it," and urged the creation of a comprehensive study—in this case, a two-tiered, long-term effort that would begin with a "reconnaissance phase," paid for by the federal government and taking a year to eighteen months to complete. This would be followed by a "feasibility phase," whose costs would be shared. The Corps apparently wanted to start over and examine all facets of river management again:

> Dovas said that in the reconnaissance phase, everything is examined, including what has gone before, work done by the state of Illinois, the environment, and hydraulics. He said there would be public hearings and preliminary cost figures given.... [T]he next phase would be the feasibility phase, and this would call for funding on a 50–50 basis by the federal government and the local sponsor [presumably the KRBC].[4]

The Corps may have been offering the KRBC an olive branch. Or perhaps it saw the moment as an opportunity to enlarge federal control over the Kankakee. Or perhaps it simply saw a need for a long-overdue comprehensive project, partly funded by the US government, for flood control and habitat preservation on the Kankakee River. Either way, the KRBC was having none of it. At a special meeting to consider the proposal, the commissioners voted 18 to 1 to reject it, and also voted down a proposal by the Indiana DNR because it suggested asking the federal government for money. The DNR's program, like the Wildlife Federation Fund's plan, was also based in part on the 1976 study and proposed construction of federally funded wide levees for flood control. Michael Aylesworth spoke strongly against any involvement by the federal government: "On principle, I'll stand opposed to involving the federal government any more than necessary," he said. "It'll run too long with too much money. We don't need any elaborate, multi-million-dollar projects. We haven't had federal funding from day one, haven't asked for any and couldn't care less."[5]

With the rejection of proposals from the Army Corps of Engineers and the Indiana DNR, the KRBC was in an awkward position for a planning organization. The commissioners had just suffered a stinging rejection of a project they had spent years planning and preparing for, and now they had voted themselves out of ideas for any kind of basin-wide management program. They needed a fresh start, and they decided that the DNR proposal, with deletion of the words "federal government," might not be so bad. At its meeting in September, the commission unanimously approved a carefully amended version of the proposal, a version that allowed only KRBC-approved agencies "to participate in the planning process."[6] The US Army Corps of Engineers had permit power over the river, but it would not be allowed a hand in charting its future.

The future, in fact, was very much on the minds of the DNR officials who drafted the proposal, which they called *The Kankakee River—A Program for the Future*. The title placed it in the ideological realm of the proposals of the Corps and the National

Wildlife Fund: it was a long-term program and it included more than flood control. It is worth examining this paper in some detail because it was clearly intended to provide a point of departure from Indiana's previous, mostly unsuccessful, attempts at Kankakee River management. The proposal acknowledged historical reality but contained modern ideas about the river and the marsh. The lead of *The Times* article about the program included this introduction: "In the past, the commission emphasized flood control and drainage on the river. In the future, the river will be considered multi-functional—including agricultural, environmental and recreational uses." The article goes on:

> "The days of the original Kankakee and its vast marshes and wetlands are forever gone," the IDNR plan says. "It can be said with equal finality that the era when it was public policy to encourage the reclamation of the swamp and overflowed land and the installation of a massive system of drainage channels and works is likewise forever gone."[7]

The DNR's paper was not so much a "plan" as it was a framework for creating a plan—perhaps several plans—for flood control, erosion control, wildlife management, and recreational use. The paper urged the recognition of several "present realities" that must be considered when planning for river management, including the reality of Illinois's opposition to any project that would threaten increased river flow. The paper argued for a view of the river that accommodated agricultural as well as conservationist concerns:

> The best interests of the people of Indiana dictate that the river serve as the basin outlet for floodwater runoff and drainage; for instream uses such as water supply, boating and recreational uses; and the remaining wetlands should remain in the state for fish and wildlife, public use and floodwater storage.

Some realities were plainly incompatible. The paper acknowledged that "as it exists today, the river is inadequate to serve the desires and needs of the agricultural sector and some urban communities with respect to flood protection and drainage." In other words, flooding of farmland and residential areas like Sumava Resorts and parts of Shelby was inevitable given the narrow, shallow channel of the channelized Kankakee River. On the other hand, the paper pointed out that "a substantial enlargement of the capacity of the main channel of the river is not in the best interest of the people of Illinois or Indiana, either from the standpoint of flood control or from environmental considerations." In other words, that shallow, narrow channel was not going to be enlarged to carry more water away from Indiana lands. Illinois was not going to allow it.

The Army Corps of Engineers most likely would not permit it, and the Indiana DNR would reject any effort that threatened wildlife.[8]

The study gave two alternatives for the future. The first was simply to "muddle along" as in the past. This option was an accurate summary of actions that had historically failed to achieve lasting results, which involved

> [c]ontinued efforts at levee repairs; occasional private efforts at dredging and levee construction, the continued "uncoordinated" private development of pumping plants for tributary damage, the continued "inexorable" loss of wetland areas at various locations; and continued dissatisfaction and controversy on all sides and failure to achieve expectations by any interest.[9]

The second option was hardly new. Illinois officials had suggested it in 1978, but now it was articulated by the Indiana DNR and endorsed by the KRBC. Presumably it could be taken for Indiana's official position, at least for the moment: "[A]ll interests involved [should] reach a consensus on, and make a conscious decision and commitment to support, a plan which holds substantial promise serving the multiple uses of the river."[10]

The language of this option belies the conviction of its intent. All "interests" could certainly agree that reaching a consensus was important (Indiana and Illinois had already agreed to this, in principle), but what consensus was possible? Calling for a "conscious" decision and commitment is an empty rhetorical gesture, for how could either be otherwise? The DNR was groping for language that would convince parties with irreconcilable beliefs to work together for common solutions. It was not easy to find such language, but they were trying.

The DNR suggested an "implementation plan" for the second option, and outlined several points that a such plan should follow. Among them was to seek funding from an appropriate agency, and to designate the KRBC and DNR as planning sponsors. In addition, the KRBC should include environmentalists and, "hopefully," the state of Illinois in a planning advisory committee. The clearest and most concrete suggestions had to do with flood control projects for the short term, such as construction of wide levees (more about this below) and clearing of significant logjams on a case-by-case basis, carrying out "flood fighting" efforts as needed, and performing emergency repairs.[11]

The KRBC was to follow some of these suggestions, incorporating them into long-range plans, as discussed later in this chapter. In the short term, the commissioners had been mulling over a levee project for some time, and even had funding lined up. This project was to expand into the KRBC's signature effort in flood management for several years.

The wide-levee project began with a plan to repair and reinforce an old levee in Lake County. The state had approved $360,000 for the project, and the commission had hopes that it could be the start of major work throughout the Kankakee River basin.[12] Construction could not begin, however, without certain information about the river. This required the KRBC's bête noir, a study. Not just any study, either. This one was to be funded by the US Soil Conservation Service. And although the SCS was part of the federal government, it was presumably a "KRBC-approved" source of federal funding, unlike the US Army Corps of Engineers, which decidedly was not. The criticism of "studies" by Indiana farmers and the KRBC was understandable. As this history shows, the river had been the subject of many studies since at least 1859, when John C. Walker had financed his own report about the river (see chapter 3). Indeed, a comprehensive list of Kankakee River studies *alone* could fill a respectably-sized book. But those who wanted quick answers to the problems of the Kankakee River in the twentieth century ignored a salient, inescapable fact: channelizing eighty miles of river and draining a huge marsh had created a half-million-acre flood plain that *by its nature* was incompatible with large areas of human habitation and agricultural land use. That massive environmental insult had created problems of massive complexity—the existence of so many studies spoke eloquently of the immense volume of what humans had to learn, not only in order to drain the marsh, but what they had to know in order to manage the consequences of that environmental insult. Draining the marsh had been the easy part. Those who believed, as one farmer put it, that "the only solution for flooding is cleaning four to five feet of silt from the river bottom"[13] had little real understanding of the land they lived on. River channels like the Kankakee's, composed entirely of sand, cannot be dredged enough. The sand is always there; it always returns.[14]

The KRBC understood that in order to make the levee project work, engineers had to know something about the hydraulics of the river in the area where the levee was to be built. According to Jody Melton, project director of the KRBC, this knowledge could be provided by a study of high and low points of river flow. The US Soil and Conservation Service had offered $250,000 for this study. Other studies were also on the agenda. Melton pointed out that the commission had set aside $580,000 to pay for aerial photography and mapping of the river. Melton was referring to studies needed for what the KRBC was now calling "the wide levee program," which Indiana officials hoped would be a long-term solution to flooding problems. Melton cautioned that such studies alone could take "three or four years to complete."[15]

With this project, the KRBC was moving with all deliberate and appropriate speed. Levee construction would need a permit from the Army Corps of Engineers. KRBC members understood that such a permit would require studies that contained sound engineering reasons supporting the project. There had been many studies of the Kankakee. There would be many more. A river is a complex environmental phenomenon in its natural state. A damaged river, inhabiting a vast flood plain, is a beast.

By May 1984, the KRBC had its permit for levee construction from the Corps, and Jody Melton announced that bids for the work would be taken in the summer. Melton's increasing exposure as commission spokesman was a sign of a subtle shift in the public stance of the KRBC. Formerly, the chairman, or president, who was elected from among the twenty-four commissioners, was routinely sought out for information about KRBC activities. Now Melton, executive director and paid staff member, had become the commission's professional contact person. He would hold the position for over thirty years, lending the KRBC much-needed consistency and credibility as an official state organization responsible for Kankakee River management.

In October, the KRBC announced that it would ask the Indiana legislature for $950,000 to acquire land for wide-levee projects.[16] The plan was to build levees some distance from the river and allow the land between levees and the river to flood during times of high water. The concept of wide levees was simple: the idea was to contain flood waters in a wide channel that would not allow a swift current to develop, thus mitigating destructive overflow into farmland and towns. In effect, excess water would be impounded—imitating the natural effects of the marsh itself, which had absorbed and held water. Engineers William Whitten and John Campbell had recognized this essential property of marshland in 1892, when they had warned against channelizing the Kankakee, as discussed in chapter 7. Illinois officials had proposed a similar strategy in 1976, advocating for a series of "buffer areas" along the river to catch runoff and hold water back from the river.[17]

The idea caught on. The first meeting of the KRBC's Wide Levee Planning Committee was held in Valparaiso in January 1985. A reporter from *The Regional News* was impressed, saying the "meeting could go down in history as 'a first'":

> It was a first in that representatives of the state of Indiana, the state of Illinois, a number of conservation and wildlife groups and several state and area governmental bodies and organizations got together and seemed to be in total agreement on something—wide levees are the way to go in flood control on the Kankakee River.[18]

The wide-levee concept resurrected ideas from the Soil Conservation Services study of 1976, combining flood control, recreation areas, and wildlife refuges. In addition to a project in Lake County, for which permits had already been secured, the KRBC was considering a site in LaPorte County as a possible location for the park–levee combination. Even the General Assembly seemed to like the idea. Michael Aylesworth, former chair of the KRBC, reported that the legislature was looking favorably on his request for funding of projects in both Lake and LaPorte counties.[19]

That year, the river reminded everyone of the necessity of a consensus for flood control. In January, the chair of the KRBC warned that severe flooding was expected.[20] Heavy rainfall and a rapid winter thaw released flood waters along the Kankakee from South Bend to the state line late in February, and though dikes in Lake County held, Shelby was threatened by high water. West of the state line, the river overflowed in Wilmington, leaving only rooftops showing in some areas.[21] Water continued to rise, and by March, a dike broke not far from I-65 near Shelby, and within days flood water was a half-mile from the town.[22] Farmland was still underwater over a month later in Porter County, and residents there were still waiting for emergency levee repairs.[23]

Hope for wide levees was high. Both the DNR and the KRBC were optimistic about a basin-wide project that would include levees and wildlife habitat. Reconstruction of the Williams Levee in Lake County, begun in July, was intended to be the "first phase" of such a project, and according to DNR director James Ridenour, the state was "committed to the concept." If the Williams Levee was successful, locations for more in other counties were under consideration, and funding was in place for long-range planning. KRBC director Jody Melton said that the computer model funded by the Soil and Conservation Service could be used to build maps showing the best locations for setback levees along the river.[24]

Individual counties began to buy into the concept of wide-levee areas with parks. In 1986, the KRBC accepted proposals from park boards in St. Joseph, LaPorte, Porter, and Lake Counties to begin open-space planning for recreational use.[25] A year later, the commission adopted a proposal from LaPorte County to construct a wide levee at the confluence of the Kankakee and Yellow rivers.[26] That same year, commissioners authorized funds for a similar project in Porter County.[27]

By the end of the year, however, momentum for the project stalled. The legislature had appropriated $1.8 million for the KRBC, but in November, the State Budget Committee refused to release the money unless the KRBC produced a master plan for wide-levee projects throughout the basin.[28] As noted above, signals from the legislature for the wide-levee plan had been favorable. Conditions seemed right for a new, comprehensive proposal. The request for a master plan presented the KRBC with an opportunity to launch its most ambitious plan yet for river management. The following March, commissioners signed a contract with SEG Engineers of Indianapolis for a plan that covered flood control, drainage, recreation, wetland preservation, water quality and supply, and wildlife habitat preservation.[29]

As the generation of such wide-ranging studies go, the SEG Master Plan took relatively little time. Although the December 1988 deadline slipped, and printing problems caused another delay, the KRBC had in hand a completed document ready for distribution by May 1989.[30] However, many landowners along the river did not have to read the actual master plan to know what they thought of it. Opposition from Kankakee farmers, sometimes fierce, grew in strength throughout 1988.

Grassroots criticism began early that year. Officials in Porter County, where the KRBC had tentatively sited a wide-levee project, objected to its cost, the fact that existing levees would have to be destroyed, and were concerned about ownership of land within the levees.[31] Such questions about details were routinely handled by Jody Melton and others, but in July, farmers turned out in force at the first public information session on the master plan. They were nearly unanimously opposed to the project.[32]

Some comments were predictable, echoing 1978 criticisms of prioritizing environmental concerns over agricultural ones. One farmer charged that "the engineers say they don't want to drive the wildlife out of that area. So instead they want to flood all this private property and drive us out. This project is crazy." Some believed that the project would remove "valuable and productive farmland from private control" without offering any chance of flood improvement. A farmer from North Judson charged that the whole idea was "insane," and offered a simple alternative: "What you need to do is clean out the river, take the dredged dirt to build up the banks and we'll be OK. I don't think you know what you are doing," he said to the KRBC members. Dredging was a popular idea. Another agreed that this was the best method of flood control: "We need to make the river deeper, not wider," he said. Not every farmer at the meeting spoke only for parochial interests. One man spoke for the river, demonstrating that farming and concern for the environment were not necessarily and always incompatible. He thought the levee project should be expanded, with more land allotted to the areas between levees and the river. "Since I was a child I've prayed for something to be done to help the river," he said. He seemed to understand what restoration of the marsh would accomplish. "I would gladly sacrifice my land for 60 miles of beauty," he added.[33]

The idea of sacrificing any land for the sake of beauty, or for any reason having to do with restoration, was vigorously rejected by a great majority of landowners. By November, they were organizing. As in 1869–71, when basin citizens protested against the Kankakee Valley Draining Company, emotions—and rhetoric—ran high. Allen Chesak, treasurer of the newly-formed Citizens for Preservation of the Kankakee River basin, issued an "open letter" to "all citizens of the Kankakee River watershed." The letter claimed that wide levees would do little to mitigate flooding and would be catastrophic for the basin: "Poor drainage results in no jobs, no industry, no farms, no schools, no pollutions control, and no future. This seems to be what the KRBC wants." Chesak conceded that the commission had done "some good," but now it had "changed its way of thinking to that of the Department of Natural Resources." Chesak charged that the DNR wanted to take over all lands along the river and use them for its own purposes.[34]

Later that month at the KRBC's meeting in Valparaiso, two hundred people showed up, most protesting the wide-levee plan. Harold Hodges, president of Citizens for the Preservation of the Kankakee Basin, presented a petition opposing the levees signed by two thousand Kankakee basin residents. Again, the DNR came in for heavy criticism, one speaker claiming that it was "the biggest obstruction the commission [the KRBC] faced." Hodges insisted there was a simple answer to the problem of flooding: "Clean the river," he said, apparently ignoring the fact that the US Army Corps of Engineers, not the KRBC, had halted that project.[35]

Representatives from environmental groups also attended the meeting and expressed their support for wide levees. They liked them because they provided for restoration of the marsh in levee areas. Illinois did not raise objections to the plan because engineers were satisfied that it posed no threat to the river past the state line.[36] As mentioned earlier, the wide-levee idea featured aspects of flood control long advocated by Illinois officials. In Indiana, however, the weight of public opinion began to press on state lawmakers, and on the KRBC.

Early in 1989, it was clear that support for the KRBC in the General Assembly was weakening when one of the KRBC's chief supporters took aim at the master plan, which had not yet been published. Representative Edward Cook, who had co-sponsored the bill to create the KRBC in 1977, called the plan "unworkable" and too expensive.[37] In February, a delegation from Citizens for Preservation of the Kankakee River Basin presented legislators and the lieutenant governor with a petition opposing wide levees.[38] In May, the delayed master plan report was finally available and made public, but the KRBC was in no hurry to adopt it. In July, citing concerns raised by Citizens for the Preservation of the Kankakee Basin, the commissioners appointed committees to examine specific issues in the plan.[39] The following month, Jim Jontz, a member of a House committee funding part of the master plan, traveled to northern Indiana to listen to citizen complaints. He was sympathetic.[40] By the end of the year, the master plan, which had been ordered by the General Assembly to justify funding for wide levees, was officially in limbo. The legislature allotted $1.5 million to improve the Brown Levee in Lake County, but there was no authorization of money for a levee system throughout the Kankakee basin.[41]

In 1990, political support for the KRBC master plan, and the wide-levee concept, became toxic as opposition to the plan became a major campaign issue. In February, Michael Aylesworth, former president of the KRBC, and former proponent of wide levees, decided to challenge Walter Roorda, co-sponsor of the bill to create the KRBC and long-time representative from the Sixteenth district, for his House seat. Aylesworth made it clear that opposition to the master plan was high on his political agenda:

> "The project [wide levees] is one of such grossly gigantic proportion that it cannot ever be remotely justified in economic reality," said Aylsworth, a farmer and a ten-year member of the Kankakee Valley Commission. "Not only are the initial construction costs outrageous, but the annual maintenance costs of electricity and manpower would be exorbitant. If we want to protect our wetlands, the state of Indiana should buy the land and put it into protected reserves."[42]

Both Roorda and Aylesworth were Republicans. Roorda was also opposed to the master plan, but Aylesworth was more opposed. He said that "he entered the race because he did not see any leadership coming from the incumbent concerning opposition to the wide levee system."[43] Aylesworth lost the primary election, but not by much. Roorda squeaked by with 136 more votes.[44] As far as the master plan was concerned, it hardly mattered. The plan languished throughout 1990. The KRBC did not officially adopt it, and the legislature continued to ignore it.

By 1991 the KRBC had not received funding since 1987, and there was no funding that year.[45] The master plan expired. Slowly, it simply faded from public attention. Unlike the clearing and snagging plan, it died a quiet, almost unnoticed death because legislators, who had demanded it, refused to support it. Some landowners on the river didn't quite understand that. In 1993, in response to a farmer's criticism of the plan, Jody Melton replied, "The commission has never adopted the wide-levee plan in its entirety. Full implementation of the plan would cost $100 million and who has $100 million these days? There is no reason to advocate its rejection because it's not going to happen." Melton went on to say that the KRBC was taking a "piecemeal approach," concentrating on building levees in Lake County where the commission had "good support."[46]

The master plan for wide levees was the last comprehensive proposal for river management that the KRBC made. The plan had promise. Environmentalists liked it, Illinois did not object to it, and the legislature, at least for a while, smiled upon it. Like the Soil and Conservation study of 1976 it was complex and expensive, and those reasons alone made it a hard sell. Like that earlier study, it was also very ambitious. The proposed system of levees would form a floodway of over 30,000 acres stretching upriver from the state line to US 30, a distance of about fifty miles. Flow through the levees from thirty-nine different tributaries—lateral ditches constructed to drain the marsh into the Kankakee—had to be managed with a complex system of pumping stations. Existing levees, consisting of the spoil banks left by dredging, had to be breached or destroyed, and several roads and bridges had to be reconstructed. Land acquisition and construction costs would be over $100 million and the plan did not provide estimates of the cost of maintaining pumping stations or the cost of electricity.[47]

High costs and complexity aside, social and political realities in Indiana made the plan impossible to implement. The master plan's own guidelines, adopted from the DNR's 1983 *Program for the Future*, articulated the conditions of its inevitable failure. Those guidelines stipulated that "all interests in the basin will have to be involved in determining the course of the action on the Kankakee."[48] Involving "all interests" apparently required the impossible task of gaining the *approval* of all interests, including residents along the Kankakee. At the beginning of the century, many residents had bitterly fought the draining of the marsh. Approval had never been unanimous. It didn't have to be. Courts, including the Indiana Supreme Court, had eventually ordered that dredging continue. At the end of the twentieth century, the state refused to bear the cost of comprehensive flood control, and no state institution would order it.

Some might say that the rejection of the wide-levee plan was a legitimate result of a democratic process: a significant number of stakeholders prevailed in a struggle for control of a natural resource. Michael Daube, farmer and chair of the KRBC, had a different idea. He believed that "a small band of farmers" had "made unsubstantiated claims, distorted the commission's intentions and disrupted meetings." He provided an example: "One guy came to a meeting with a baseball bat and slammed it on the table. That's the kind of people we're dealing with."[49] Residents not only objected to costs and the complexity of the master plan. They rejected any plan that included major restoration of the marsh.

Much of residents' bitterness had been directed at the DNR because, as one critic put it, "the DNR wanted to take over all lands along the river and use them for its own purposes."[50] These "purposes"—habitat restoration and the protection of wildlife—threatened many farmers because they believed they would be forced to give up farmland—their means for a livelihood. To many in the basin, the Kankakee was no longer a river, it was drainage ditch. For these landowners, efforts to "bring back the marsh" were bound to be met with hostility. As Allen Chesak of the Citizens for the Preservation of the Kankakee Basin put it, a proposal like the master plan, with provisions for wildlife habitat and recreation, "misses the fulcrum of the river. The river is drainage."[51]

The failure of the master plan had the benefit, at least, of clarifying a few things about flood control planning in the Kankakee basin. Like the Soil and Conservation study of 1976, the master plan showed that a comprehensive proposal to control floods on the Kankakee River was very, very expensive. Estimated costs to implement the 1976 plan were $130 million. The master plan came in at over $100 million. Jody Melton spoke for a modern consensus: no one had that kind of money to spend on fixing the problems caused by draining the marsh. It was also obvious by the mid-1990s that any plan would have to gain the approval of Kankakee landowners, who generally believed that only dredging or clearing the river could control flooding. Neither the Army Corps of Engineers, or the state of Illinois, or the Indiana DNR would approve that. What was especially alarming to conservationists, however, was the deeply intransigent antipathy to restoration shown by farmers on the river. That antipathy would take a bizarre turn for the worse before the decade was over. And it would play a pivotal role in halting comprehensive restoration planning in the marsh well into the next century.

Some small restoration projects were a success. In 1993, a coalition of local and national organizations formed the Indiana Grand Kankakee Marsh Restoration Project to restore 26,000 acres in various locations throughout the old marsh. The group was composed of the Indiana DNR, Lake County Parks, The Nature Conservancy, Waterfowl USA, and others. The group applied for a grant from the US Fish and Wildlife

Service. Even the KRBC expressed interest in donating land it owned to the project. Eventually, the Restoration Project was able to purchase a relatively modest 12,000 acres for protection, including part of the Beaver Lake basin in Newton County.[52] (This project was described in chapter 6.)

A larger project, however, was extremely controversial. The success of the Indiana Grand Marsh Restoration Project inspired the US Fish and Wildlife Service in 1997 to propose a much more ambitious project: the conversion of 30,000 acres in the basin into the Grand Kankakee National Wildlife Refuge. The service had in mind a different kind of refuge. Instead of the traditional large tract of land within a specified boundary, this would be "a series of smaller tracts" that linked existing areas of protected wildlife habitat in the Kankakee River watershed in both Illinois and Indiana.[53] Eventually, the Wildlife Service hoped to combine these areas into newer ones established by partners such as The Nature Conservancy and the Illinois and Indiana DNRs which would form a corridor of 100,000 acres of protected habitat. There were strategic reasons as well as ecological ones for this innovative plan. The Wildlife Service believed that by "generating a suite" of sites it could bypass parcels difficult to acquire because of cost or owner opposition.[54] It was a good plan for the acquisition of land for environmental protection, but it badly underestimated the politics, and the social reality, of rural northwest Indiana in the late twentieth century. Opposition—and misinformation—began immediately.

Indiana congressman Steve Buyer led the attack against the refuge. At a press conference in Plymouth, Indiana, Buyer announced that the Fish and Wildlife Service had plans to turn 100,000 acres of land in northern Indiana into marsh and wildlife refuge to prevent flooding. Buyer was concerned, he said, that plans were being made without "input" from the people who lived on that land. He was worried that the Wildlife Service would "obtain control of 100,000 acres in Indiana," and be more concerned with the population of "spotted owls" than with farmers' issues. Buyer insisted that no plans for restoration should be made before an Army Corps of Engineers feasibility study for flood control, already underway, could be completed.[55]

The Fish and Wildlife proposal, of course, was for 30,000 acres in two states; there were no plans to "control" 100,000 acres of land in Indiana. Buyer called the refuge proposal a flood control project, but it was not. It was a plan for ecological restoration. The Army Corps's feasibility study that Buyer mentioned was a basin-wide flood control study that he and two Illinois congressmen had ordered.[56] It had barely begun in April 1997 when Buyer made this announcement. State funding was uncertain, which made its completion date also uncertain. Besides, there was no reason to delay planning for the refuge because of this study. Federal lawmakers would require the Corps and the Wildlife Service to sign a memorandum of agreement to avoid any duplication of effort.[57]

Buyer's reference to the "spotted owl" no doubt was an allusion to a controversy in Oregon a few years before when a federal judge had halted logging in order to prevent possible extinction of the spotted owl, a threatened species. This owl, of course, was native to the American northwest. It was not found in Indiana. Buyer's charge that plans for the refuge were being made without local input was baseless. The Fish and Wildlife Service was obliged under the National Environmental Policy Act, like all other federal agencies, to hold public meetings and prepare review documents when environmental projects were planned. As a federal legislator, Buyer surely knew this.[58]

Buyer's campaign against the refuge included letters to the governor and other elected officials in Indiana and the US Congress, to the Fish and Wildlife Service, and press releases criticizing the plan. Common themes in them all were the size of the proposal—Buyer called it "potentially" 100,000 acres, but never mentioned the actual figure proposed, 30,000 acres—and the fact that the proposal might somehow interfere with the feasibility study being conducted by the Army Corps of Engineers. Buyer stressed his belief that the Corps's study, sponsored by himself and other congressmen, was "best suited" to examine basin-wide initiatives.[59] He also mentioned "limited" public input at the current time, even though he had been briefed by the Fish and Wildlife Service about the legal requirement for public meetings, which were scheduled for summer 1997.[60]

In light of their multiple "concerns" about the refuge proposal, all three sponsors of the Army Corps's study—Representatives Steve Buyer (Indiana), Tom Ewing (Illinois), and Jerry Weller (Illinois)—signed a letter to the chair of the Subcommittee on Interior Appropriations, attempting to forestall 1998 funding for land acquisition by the Fish and Wildlife Service.[61] The effort to deny funding would bear fruit.

In his press release, Buyer said it was troubling to find that many residents had not heard of the refuge project, and he implied unwarranted credit for making sure the Fish and Wildlife Service would solicit public input on the proposal. What is really troubling, however, is this alarmist, and false, claim:

> My office was first notified last spring that the wildlife refuge was in its initial stage and the process would take several years. In June, however, I learned that the refuge could be established by late fall of this year [1997]. The pace of this project is too quick and is unacceptable. I have serious doubts that the Fish and Wildlife Service could fully hear the concerns of local residents, and that local residents would have adequate time to make their concerns known.[62]

The press release was published in October 1997. That a 30,000-acre refuge could be "established" in literally only a matter of weeks was manifestly absurd. The service did not actually approve the plan for nearly two more years, in 1999, and only then after extending its public review period (as requested by Buyer) until June 1998.[63]

Buyer also distorted the Fish and Wildlife Service's policies on land acquisition, claiming that the service "may be quick to designate an area a wildlife refuge, but may not be as quick to provide money for land purchases, leaving residents who wish to see their land 'boxed in' and unable to put their property on the open market." In addition, he charged that the service sometimes restricted plowing or pesticide application on adjacent properties during bird migratory seasons.[64]

These charges were refuted by the Fish and Wildlife Service's own public relations campaign and others in the press, but Buyer's inflammatory tactics worked. By early 1998, landowners throughout the Kankakee basin were meeting, and agitating, against the refuge. Sometimes, paranoia about a United Nations plot to take over US farmland became conflated with criticism of the wildlife refuge proposal. In Thayer, Indiana, eighty residents from Indiana and Illinois listened to speakers who linked the project to "a subtle federal government effort to implement the terms of a United Nations agreement that [would] effectively turn over the land-management decision-making powers of hundreds of thousands of acres of United States land to the United Nations."[65] The following month in LaPorte County, an "activist" hijacked a county commissioners' meeting by falsely advertising it as a public meeting about the refuge. Two hundred people showed up, forcing the commission president to allow a lengthy discussion about the refuge. An article in the Michigan City *News Dispatch* described what happened:

> Most of the speakers did not veil their distrust of the federal government and its agency representatives or hide their belief that neither can be taken at its word. When Dave Hudak, a Fish and Wildlife Service spokesman, said only land bought from willing sellers would be acquired, and that eminent domain would not be used, harsh mutterings and open groans were audible... "They say they'll only buy land from willing sellers, but believe you me, they can make a willing seller out of just about anyone," one Starke County resident said.

The article pointed out that some in the crowd "seemed to believe that the proposed refuge is a United Nations plot to get a foothold on the land and then turn the entire basin back to its pre-European-development state." This idea was widespread. According to the article, a Michigan City television channel had featured a story about someone who "adhered to this particular theory."[66] In Wheatfield, six hundred residents sat in the aisles and on the stage to attend the first public forum about the refuge sponsored by the Fish and Wildlife Service. More than sixty people spoke, and for one observer it was obvious that the "great majority of the crowd was diametrically opposed to the refuge."[67]

The press loved the story about the UN taking over farmland, and why not? John Husar, columnist for *The Chicago Tribune*, interviewed Bill Kruse, an insurance agent from Roselawn, Indiana, who headed the Kankakee River Task Force. The motto of this group was "Keep the U.N. Out of Indiana." Kruse quoted US Congressman Steve Buyer as confirming that the UN actually ran Yellowstone National Park, Mammoth Cave, the Great Smokey Mountains National Park, and some forty-four other national parks.[68]

Kruse had circulated a flyer that stated that "national parks are under United Nations Control! This according to Congressman Steve Buyer! The entire Kankakee River's 3.3 million acres watershed will be a national park under U.N. control unless we unite!" Buyer's office was quick to deny that he had ever made such statements, saying that any involvement by the United Nations in the refuge project was "ludicrous."[69] No doubt, Buyer cannot be blamed for such mad conspiracy theories, or their circulation. But his demagoguery was inspirational.

Stories about a UN takeover of Indiana farmland may have been popular in the press, but a small, unassuming article in the *Indianapolis News* told a more realistic story about public reaction to the refuge proposal. In contrast to the angry protests at some meetings, an "overwhelming number of letters in favor" of the plan had been sent to the Fish and Wildlife Service. By June 1998, it had received about 6,320 letters, comments, and postcards supporting the refuge, compared to 865 against. Letters from residents living near the refuge area were running around 4 to 1 in favor of the plan.[70] It wasn't going to matter. Official opposition was coalescing around an issue emphasized by Congressman Buyer repeatedly: the putative conflict between the Army Corps of Engineers' bi-state feasibility study and the plan for the refuge. As early as November 1997, the Kankakee River Basin Commission adopted a resolution opposing the refuge until the Corps's study was completed.[71] The Kankakee–Iroquois Regional Planning Commission followed suit in 1998. In addition, the Indiana Farm Bureau expressed strong opposition to the project, and the Kankakee County (Illinois) Farm Bureau and the Newton County (Indiana) Board of Commissioners raised strong objections.[72]

All this might not have mattered, but what the US Congress did in 1999 mattered a lot. The spending bill for 2000 prohibited the Fish and Wildlife Service from creating a wildlife refuge in the Kankakee basin. Possible conflict with the Corps of Engineers' study was specifically cited: "None of the funds in this Act may be used to establish a new National Wildlife Refuge in the Kankakee River basin that is consistent with the U.S. Army Corps of Engineers' efforts to control flooding and siltation in that area."[73] In 2000, Buyer was successful in stopping funding for the refuge again, through 2001.[74] There was still no funding in 2002, and in that year, the Fish and Wildlife Service moved its planning office from Plymouth, Indiana, to Watseka, Illinois, because it was in an area where people supported the refuge.[75] The Indiana project was quietly dropped after 2002. The Army Corps of Engineers' bi-state feasibility study, so often cited by Representative Buyer as a reason to delay the national refuge proposal, evaporated from public view. It was never published. In 2016, the Kankakee National Wildlife Refuge and Conservation Area was established in Iroquois County, in Illinois. No national wildlife refuge has ever been created on the Kankakee River in Indiana.

In 2002 the Kankakee River Basin Commission marked its twenty-fifth anniversary. For a quarter of a century, the commissioners had worked hard at their primary task of establishing effective flood control measures on the Kankakee River. Formidable opposition from Illinois and conservationists, antipathy from its constituents (landowners on the river), and lackluster support from the Indiana General Assembly had frustrated those plans. It would require seventeen more years and a major flood before legislators recognized that the KRBC needed an overhaul, and a long overdue change in status, to ensure that Indiana could effectively meet the historic challenges of a damaged river in a new century. And though a national wildlife refuge on the Indiana Kankakee never became a reality, conservationists forged new, creative partnerships with landowners in the twenty-first century to preserve marshland.

# NOTES

1. David McCarty, "$200,000 River Basin Study Urged," *The Indianapolis News*, April 12, 1983. All newspaper articles, except as noted, were accessed online at Newspapers.com or Newspaperarchives.com.
2. Linda Schmidt, "Board Won't Fight Scrapping of Clear-Kankakee Plans," *The Vidette-Messenger* (Valparaiso, IN), April 22, 1983.
3. Richard Salomon, "Panel to Rethink Study Resolution," *South Bend Tribune* (South Bend, IN), July 15, 1983.
4. Salomon, "Panel to Rethink Study Resolution."
5. Scott Cottos, "Denial of Federal Plan Means KRBC Will Rely on 1976 River Study," *LaPorte Herald* (LaPorte, IN), July 29, 1983. From the Kankakee River Basin Commission archives (henceforth KRBC archives), Kankakee River Basin and Yellow River Basin Development Commission, Valparaiso, IN.
6. Richard Salomon, "Kankakee Report Identifies Goals," *South Bend Tribune*, September 25, 1983.
7. Diane Donovan, "Long-Range Plan Accepted," *The Times* (Munster, IN), September 23, 1983.
8. "DNR River Plan Rejected by Commission," *Southlake Register* (Cedar Lake, IN), August 3, 1983. From the KRBC archives. This article is a very good contemporary source for the details of the DNR's *The Kankakee River—A Program for the Future*.
9. "DNR River Plan Rejected by Commission," *Southlake Register*, August 3, 1983.
10. "DNR River Plan Rejected by Commission," *Southlake Register*, August 3, 1983.

11. "DNR River Plan Rejected by Commission," *Southlake Register,* August 3, 1983.
12. Linda Schmidt, "Focus Now on Levee Work," *The Vidette-Messenger,* April 22, 1983.
13. "DNR River Plan Rejected by Commission," *Southlake Register,* August 3, 1983..
14. *Kankakee River Flood and Sediment Management Work Plan* (Christopher B. Burke Engineering, 2019), 35–36.
15. "KRBC Gets $250,000 for Kankakee River Studies," *The Vidette-Messenger,* November 14, 1983.
16. Richard Salomon, "Land Acquisition Funds to be Studied," *South Bend Tribune,* October 14, 1984.
17. Jerry Morgan, "Emotions on Kankakee Run Deep," *Kankakee Journal* (Kankakee, IL), February 15, 1976.
18. Jeff Mayes, "Committee Likes Wide Levees," *The Regional News* (Porter County, IN), January 31, 1985. From the KRBC archives.
19. "Kankakee May Get New Park Along with Wide Levee System," *The Regional News* (Porter County, IN), January 31, 1985. From the KRBC archives.
20. Richard Salomon, "Severe Floods Expected in Kankakee Area," *South Bend Tribune,* January 25, 1985.
21. Melanie Csepiga, "Residents Along Kankakee Watch," *The Times* (Munster, IN), February 25, 1985.
22. "River Breaks Through Wall; Shelby Braces for Flooding," *The Times* (Munster, IN), March 11, 1985.
23. Frank Wiget, "Flooded Kankakee Residents Still Waiting for Emergency Repairs," *Post-Tribune* (Gary, IN), April 17, 1985.
24. Mary Sue Penn, "Levee Work Could set Pattern," *The Times* (Munster, IN), July 26, 1985.
25. Richard Salomon, "Wide-Levee Plan Approved," *South Bend Tribune,* June 27, 1986.
26. Richard Salomon, "Levee Project on Kankakee Gets Go-Ahead," *South Bend Tribune,* April 24, 1987.
27. Jeff Mayes, "Planning Continues for Kankakee Levees," *The Vidette-Messenger* (Valparaiso, IN), July 17, 1987.
28. "KRBC Panel to Reply to State Warning," *South Bend Tribune,* November 8, 1987.
29. Richard Salomon, "Work on KRBC Plan to Begin Month Late," *South Bend Tribune,* March 20, 1988.
30. "Kankakee Basin Plan Ready," *South Bend Tribune,* May 31, 1989.
31. Jeff Mayes, "Officials at Odds on Levee Plan," *The Vidette-Messenger,* May 12, 1988.
32. Jeff Mayes, "Farmers Reject Levee Plan," *The Vidette-Messenger,* July 12, 1988.
33. Mayes, "Farmers Reject Levee Plan."
34. Richard Salomon, "Citizens Oppose Wide Levee Project," *South Bend Tribune,* November 6, 1988.

35. Richard Salomon, "Levee Foes Dominate Meeting," *South Bend Tribune*, November 18, 1988.
36. Richard Salomon, "Porter Turned Down on Funding Request," *South Bend Tribune*, April 26, 1991.
37. Richard Salomon, "Cook Criticizes Cost of Levee Plan," *South Bend Tribune*, January 15, 1989.
38. Katherine Bieker, "Legislators Hear Concerns About Wide Levee System," *The Vidette Messenger*, February 4, 1989.
39. Richard Salomon, "Committees to Get Parts of River Plan," *South Bend Tribune*, July 28, 1989.
40. Katherine Bieker, "Jontz Schedules Meeting to Talk About Kankakee River Master Plan," *The Vidette Messenger*, August 14, 1989.
41. Jeff Walz, "Brown Levee Plan Funded," *The Vidette Messenger*, December 5, 1989.
42. Jeff Walz, "Aylesworth to Challenge Roorda," *The Vidette Messenger*, February 28, 1990.
43. Richard Salomon, "Ex-KRBC Official Opposes Plan," *South Bend Tribune*, March 18, 1990.
44. "Republican's Victory Cheers Democrats," *The Courier-Journal* (Louisville, KY), May 10, 1990.
45. Richard Salomon, "KRBC to Make Do with '87 Leftovers," *South Bend Tribune*, July 7, 1991.
46. Wayne Falda, "Farmers Along the River Criticize Cost, Practicality," *South Bend Tribune*, February 7, 1993.
47. SEG Engineers and Consultants, *Kankakee River Master Plan: A Guide for Flood Control and Land Use Alternatives in Indiana* (SEG Engineers and Consultants, 1989), v, 45–55.
48. SEG Engineers and Consultants, *Kankakee River Master Plan*, v, 45–55.
49. Wayne Falda, "Farmers Along the River Criticize Cost, Practicality."
50. Richard Salomon, "Citizens Oppose Wide Levee Project," *South Bend Tribune*, November 6, 1988.
51. Richard Salomon, "Wide-Levee Project Looks Tough to Sell," *South Bend Tribune*, June 18, 1989.
52. Martha Rasche, "KRBC Eyes Flood Control Tie in the Marsh Project," *South Bend Tribune*, October 15, 1993.
53. "The Proposed Grand Kankakee Marsh National Wildlife Refuge" (Flyer, Department of the Interior—US Fish and Wildlife Service, June 15, 1997). From the KRBC Archives.
54. "Grand Kankakee Marsh National Wildlife Refuge: Preliminary Project Proposal Summary" (US Fish and Wildlife Service, 1996). From the KRBC Archives.

55. Linda L. Mullen, "Public Input First: Buyer," *South Bend Tribune*, April 22, 1997.
56. "Owners Voice Concern Over Kankakee Study," *South Bend Tribune*, July 26, 2001.
57. "Owners Voice Concern Over Kankakee Study," *South Bend Tribune*, July 26, 2001.
58. The Acting Regional Director of the Service reminded Buyer of this in a letter dated May 18, 1997. Letter to Steve Buyer from Thomas J. Kerze, Acting Regional Director, Fish and Wildlife Service. From the KRBC Archives.
59. Letter to William Alexa, Indiana State Senator and others from Steve Buyer, US Representative from Indiana, April 1997. From the KRBC Archives.
60. Reply to Buyer's letter from Thomas J. Kerze, Acting Regional Director of the Fish and Wildlife Service, May 18, 1997. From the KRBC Archives.
61. Letter to Ralph Regula, Chair, Subcommittee on Interior Appropriations, May 14, 1997. From the KRBC Archives.
62. Steve Buyer, "Kankakee Marsh Project Moving too Fast," *South Bend Tribune*, October 1, 1997.
63. "Federal Officials Authorize Creation of Wildlife Refuge," *Indianapolis Star*, August 26, 1999, and "Review Period Extended," *Rushville Republican* (Rushville, IN), April 24, 1998.
64. Steve Buyer, "Kankakee Project Moving too Fast."
65. Melanie Csepiga, "Groups Urge Fight Against Wildlife Refuge," *The Times* (Munster, IN), March 22, 1998.
66. Nicole Duran, "Wildlife Proposal Angers River-Area Residents," *The News-Dispatch* (Michigan City, IN), April 19, 1998.
67. Robert G. Ax, "Letter to the Editor," *South Bend Tribune*, June 6, 1998.
68. John Husar, "UN 'Land-Grab Plot' Scary or Silly?," *Chicago Tribune*, April 5, 1998.
69. Mike Copher, Press Secretary to Steve Buyer, in "Letter to the editor," *South Bend Tribune*, April 9, 1998.
70. "Kankakee Marsh Debate," *The Indianapolis News*, July 25, 1998.
71. Resolution 97-1, "A Resolution Opposing the Creation of a National Wildlife Refuge in the Kankakee Basin until the Army Corps of Engineers Completes the Bi-State Feasibility Study," Kankakee River Basin Commission, November 20, 1997. From the KRBC Archives.
72. Documents from these organizations courtesy of the KRBC archives.
73. "Congress Delays Funding for Kankakee Wildlife Refuge," *South Bend Tribune*, November 24, 1999.
74. Terry Turner and Linda Mullen, "Kankakee Land Buy Delayed for 2nd Year," *South Bend Tribune*, October 8, 2000.
75. "Wildlife Refuge Plan Focuses on Illinois," *The Times-Mail* (Bedford, IN), January 21, 2002.

## 11

# THE KANKAKEE BASIN IN THE 21ST CENTURY

## Lessons in Changing Values

IN FEBRUARY 2018, HEAVY RAINS CAUSED THE WORST FLOODING IN NORTH-west Indiana since 1954. Flood warnings for most of the counties along the Kankakee River were issued, and dozens of roads were closed in Lake, Porter, and LaPorte counties. Parking lots became ponds and cars were abandoned on flooded roadways. Water rose above bridges, and levees in Lake County were breached, causing evacuations from residential areas as thousands of acres of farmland flooded.[1] In some areas, floodwater persisted for six weeks. Several locations on the Kankakee River reported the highest flood levels ever recorded.[2]

There were those in the government who saw an opportunity in the crisis. The severe floods inspired calls for change from the Indiana General Assembly as soon as the new legislative session was opened in 2019. Representative Steve Gutwein believed that the KRBC could do a better job of flood management if it had fewer commissioners. "There's too many people on that commission," he said. "They don't get anything done."[3] Gutwein's proposal was to reduce membership on the commission from twenty-four (three members per Kankakee River border county) to nine (one member per county, plus another appointed by the governor). The real innovation, however, was that each member would be required to have practical experience for the job, such as knowledge of at least one of the following: construction, project management, flood control, or drainage. Requiring such experience was an important step, long overdue, in modernizing the KRBC. However, the new executive director of the KRBC, Scott Pelath (Jody Melton had retired in 2018), pointed out the flaw in the proposal: "If we go from 24 people down to nine people and there's no resources to do the major work that needs to be done, then it's just a smaller group of people having coffee," he said.[4] Pelath was commenting on the fact that the KRBC's activities had been hamstrung for the entire forty-two years of its existence by the lack of a real budget. Every project had to be funded by the legislature directly, and as noted in the last chapter, there were years when KRBC flood control projects were not funded at all.

Remarkably, the 2019 legislature agreed with Pelath. The final bill not only reconstituted the KRBC, it granted new powers, and provided a sharply increased budget. Membership was decreased to nine, and in a significant departure from the earlier commission, two advisory members from Illinois were added. This was a major step, also long overdue, in recognizing Illinois's role in consulting with Indiana on managing the Kankakee River. The law granted the commission specific powers, such as land acquisition for maintenance roads, levees, and flood storage. The commission could construct sand traps and remove sediment, could reconstruct the river channel, and was granted "other flood control actions considered necessary."[5] That year, the legislature authorized $2.3 million to begin flood management and erosion control work. Beginning in 2021, annual assessments from the eight counties bordering the river were required to generate $3 million a year for flood control projects. The commission was rechristened the "Kankakee River Basin and Yellow River Basin Development Commission" to reflect its broadened responsibilities.[6] The name was awkward, but the new, streamlined organization was not. Finally, after forty-two years, the commission had acquired the status of a proper government agency: it had specific duties and an annual budget, one that was large enough to allow for long-range planning.

That planning began immediately. A few years before, the KRBC, with funding from both Indiana and Illinois, had commissioned a new study of the river, specifically designed as a long-range plan for managing what the commission accurately called "over a century of mounting flood and erosion problems."[7] The *Kankakee River and Sediment Work Plan* was different from earlier flood management plans for the Kankakee. Unlike either the US Soil Conservation Service report of 1976, or KRBC's Master Plan of 1987, the report included information about the Illinois portion of the Kankakee, and recognized major tributaries—the Yellow River in Indiana and the Iroquois River in Illinois—as contributors to serious sand problems in the Kankakee. Instead of recommending an expensive, comprehensive flood control approach like the earlier studies, the Burke Report (named for the engineering company that created it), was, as its title promised, a true *work* plan. Its flood control approach was much more modest, much less expensive, and much more practical than earlier studies.

Rather than a river-wide flood control system, the report recommended a "strategic" approach. This meant phasing in improvements over a long period—forty years, in fact—which had the advantage of keeping yearly costs within budget. Such improvements included bank reconstruction on both the Kankakee and Yellow rivers, large wood removal, berm maintenance, logjam management, and bridge removal.

The report made no claim to solving flooding problems on the river. On the contrary, it noted that "[d]espite the recommended flood protection/prevention measures, the risk of flooding will persist for most of the river corridor; the development of flood response and flood resilience plans is recommended to further mitigate the risk to communities and landowners."[8] The report understood what was in store for the river and for residents of the basin. It offered this sober advice about flooding:

> Recognizing the extent of the existing risks and the likely future vulnerabilities in the face of a changing climate, addressing the flooding and sedimentation issues within the Kankakee River system will require both adaptation and mitigation. *Adaptation and learning how to live with floods will be necessary because there are no feasible structural solutions to eliminate the vulnerability to flooding along the Kankakee River, especially given the increasing trends in peak flows and volumes* [emphasis added].[9]

The report noted that as "extensive as the 2018 flooding was, it [did] not represent the *worst* conditions that are likely to occur along the Kankakee and its tributaries (emphasis added)."[10] Climate change was expected to increase Kankakee floods. In the future, flood response would have to be adaptive. Flooding was not only inevitable; it could never be fully controlled. However, with careful management, damage from floods could be lessened. This was a first in the history of efforts to control the "monumental problems" on the Kankakee River: flooding was officially recognized as a permanent condition.

As mentioned earlier, the study identified both the Iroquois and the Yellow Rivers as sources of sand and silt washing into Illinois. Using the recommendations of the Burke Report, the Kankakee River and Yellow River Basin Development Commission began a three-phase project to halt sediment erosion on the Yellow River. By December 2023, the commission had completed Phase Two, which reconstructed eroding riverbanks along three miles of the Yellow River. This work encompassed reducing bank grades, increasing channel capacity, reinforcing banks, and reseeding them with native erosion control vegetation.[11] The long-term work plan also identified the creation of areas for the storage of excess water as a priority. Impounding flood water was a goal of previous studies, especially the KRBC's wide-levee plan in the 1980s. In 2023, the commission, in partnership with the DNR, developed over four hundred acres of marginal farmland in Newton County to absorb excess water during flooding.[12]

With the notable reduction of sand in the Indiana portion of the Kankakee River, tensions with Illinois were greatly eased.[13] Meetings of landowners from both states had been amicable for years, in fact, although old animosities were not entirely forgotten. In 2017, an Indiana resident remembered attending meetings about the river in the 1970s armed with a baseball bat.[14] (People were still carrying bats to meetings in the 1990s—see chapter 10.) Other old animosities were also laid to rest. In 2023, the commission reported that the US Army Corps of Engineers was working on a "formal, coordinated flood response plan" for *all* of the Kankakee River. Over seventy people from both states, landowners and officials, had attended a series of workshops to gather advice for the plan.[15]

In 1983, after its permit to clear and snag the river was rejected by the Corps, the KRBC had publicly dismissed the need for more studies of the river. (See chapter 9.) In 2023, both groups were back at the table. History may or may not have much to teach us (more about this later), but it does show that studies of the river, and meetings to discuss them, will always be part of life in the Kankakee basin. The bitter disputes between Indiana and Illinois, which lasted for decades, eased as policies to manage the river have become more inclusive, that is, more democratic. Beginning in 1893, Indiana had embarked on an aggressive campaign to direct the waters of the Kankakee River as it saw fit, a campaign that was essentially proprietary, since the state was exercising its "right" to use the river to drain its lands. Over a hundred years later, the government had entered a new phase in which, finally, a kind of Jeffersonian compromise with Illinois seemed possible.[16] It was a long and costly journey.

The Burke Report's incremental, long-range approach to river management evolved over many years, and from many studies of the Kankakee. The revised policies of Indiana toward river management and relations with Illinois required a century to develop. As noted in chapter 10, draining the marsh was the easy part. Indiana found that managing the consequences of massive geomorphological change required a long and steep learning curve. Perhaps we can agree with Aldo Leopold's observation, contained in the epigraph for this book, that the law of diminishing returns is hard at work in the operations of what some are still pleased to call "progress."

In 1892, John L. Campbell pointed out the folly of channelizing rivers. He called it "vandal engineering," and pronounced that it would only "end in defeat." Campbell undoubtedly meant, at the least, that artificially changing the course of a river would not reliably produce beneficial results. (This has been recognized in modern times. See chapter 7.) He also warned that draining wetlands caused flooding. He had seen this himself after reclamation projects in other parts of Indiana had destroyed the water retention ability of marshes.[17] William Whitten discussed with approval Campbell's critiques of channelization and draining at length during a public meeting of the Indiana Engineering Society, and offered his own plan for draining that did not include channelizing the river. His audience was composed of individuals who were in charge of public works projects throughout the state. Interest in attempts to drain the marsh was very high, judging from the fact that papers on the subject were on the agendas of several of the society's meetings in the 1890s. If any in Whitten's audience were involved with plans to channelize the river a few years later, we must regret that his discussion did not resonate strongly enough with them.

As noted in chapter 7, modern river engineering has learned from past mistakes. It has learned from damaged ecologies along altered water courses, the silted-up reservoirs behind dams, and the disfigured hydrologies of drained wetlands. That history—of success and failure—is written in land and water. The story of Florida's Kissimmee River, restored after it was channelized, shows how much can be learned, and how much restored, by an honest assessment of past failures (see chapter 7). We may ask what can be learned by a study such as the present one—a work of environmental history that consists of mere words.

An important reason for a history of the destruction of the Kankakee Marsh and Beaver Lake is to tell a story that many people do not know. In her research for a PhD dissertation in anthropology (2017), Elizabeth Maree Hare spent months in northwest Indiana interviewing its residents. The subject of her work was conservation policy in the Midwest, and she focused on the communities in the Kankakee basin for her ethnographic fieldwork. She found that most people knew very little about the area where they lived:

> Virtually everyone I talked to about the wetlands began by telling me about how they came to learn of the marsh. Most didn't know about it until they were adults, even though they had spent their entire lives in the area. Much of the information about the history of the river is shared person to person, in conversation. There were a few books written about the draining of the wetlands in the early twentieth century, but very little was published after 1950. [Some of these works are explored in chapter 1.] The history has not been taught in schools, so far as anyone could recall. Several people I spoke to speculated that this was because of the opposition to the drainage back in the day, while others argued that the silence was due to the utterly unremarkable nature of the event. Draining wetlands to create farmland was simply what was done back then. People I spoke with told me that they knew that the Kankakee river had been ditched and straightened, but in an area so full of drainage ditches, that fact alone is utterly unremarkable. When recounting their personal discovery of the marsh, some spoke with a sense of wonder, others with a sadness akin to a personal loss.[18]

It is not surprising that most people growing up on the old marsh hardly knew of its existence. The few physical reminders that have been preserved are small and scattered. Who could know of the *magnitude*, much less the biotic plenitude, those fragments represent? What people see—what we all experience of any built environment—is an environment whose existence seems inevitable because our existence without it is unimaginable. It is also hardly surprising that the history of the destruction of Beaver Lake and the Kankakee Marsh is not taught in school. In order to teach the history of absence, one must first recognize that what the built environment has replaced is important, and should be *missed*. A new landform—a vast domain of farmland—does not merely "appear." The old landform—in this case a vast marsh—must first be destroyed. Few want to acknowledge that a transformation on this scale is an act of environmental violence.

As this history shows, protracted, vigorous opposition to draining the marsh consumed the citizens of the Kankakee basin for many years. That this opposition was not based on conservationist principles hardly matters—the opposition was sometimes organized, always intense, and lasted decades. That fact alone shows that draining the marsh was hardly "unremarkable." The story of such opposition ought to be told. The destruction of the marsh was not inevitable. Those who advocated for it and enabled it had to fight hard for the privilege. This salient fact matters and deserves its history: determined resistance was eventually overcome by a strenuous persistence that, had it been less dedicated, would have probably failed. More than a stunning natural area was destroyed. A way of life that many Kankakee landowners treasured and fought for years in court to preserve was also destroyed. It is not impossible to imagine what the Kankakee basin might be like today had those landowners prevailed: perhaps a vital agricultural economy could have existed side by side—literally—with a river that was not deformed by channelization. And perhaps much more of the old marsh would have been spared. We should keep in mind C. F. Rupel's reply to Dixon Place in 1897. Place, who campaigned successfully for the first major project to straighten the river, said he wanted to drain the marsh in order to create the "garden spot of Indiana." Rupel replied that we should "keep the garden spots we already have," instead (chapter 5). He spoke for many who farmed in the Kankakee basin and believed—correctly—that the gardens they had created were endangered by Place's project. They saw no reason to transform a river into a drainage ditch.

One of the residents of the Kankakee area interviewed by Elizabeth Hare for her dissertation was Pat Wisnieski, who, along with Jeff Manes, Brian Kallies, and Tom Desch, produced a PBS documentary in 2012 called *Everglades of the North: The Story of the Grand Kankakee Marsh*. The film provides an overview of the marsh's destruction, but it is more than a requiem for the loss of something grand—it is an elegy for a felt absence.

Like so many others, Wisniewski had grown up on the old marsh and only learned about it as an adult.[19] The production of *Everglades of the North* is mute testimony to the power of the marsh—a thing so absent from the landscape that most do not even know it ever existed—to inspire an imaginative presence, a recreation through film that has inspired many to learn more about the marsh. The history of absence is not just about something lost—which perhaps can be found, or "restored," to use a conservationist word for the recovery of a natural landscape. The history of absence is inspired by the feeling of something missing that ought to be present, but has been removed abruptly—*taken*, actually. The Kankakee Marsh and Beaver Lake were unique and are irrecoverable. Now they are absent from the earth. It is hardly surprising that this fact inspires a sadness "akin to a feeling of personal loss," as Hare puts it.

Images of wetland plants and wildlife in *Everglades of the North*—indeed its overall tone—are meant to evoke a sense that the marsh had value in itself—that it was beautiful, rare, and precious, and should have been preserved. Clearly, the film's creation by individuals who grew up in the Kankakee basin shows that some residents, at least, certainly do not share the values of those of an earlier time who deformed the river and destroyed the marsh. And it should be pointed out that the creators of the film were not the "fish and wildlife people" so often criticized by those who fought bitterly against projects to restore portions of the marsh. They were not "outsiders." They were not employees of the DNR, and they were not a well-financed film crew hired by a national conservation organization. They were not advocating for a national wildlife refuge on the marsh, though they certainly might have welcomed its creation. They were ordinary citizens of the Kankakee basin who shared a sense that something of inestimable value had been destroyed. The nature portrayed in *Everglades of the North* has a value all its own—an intrinsic value that transcends the economic or merely utilitarian value placed on the marsh by those who transformed it into farmland.

A history of the draining the Kankakee Marsh and Beaver Lake should make us ponder the question of our relationship to nature. This consideration might begin with an examination of the destructive power of values once held in high esteem. These are the values of an unquestioned and unreflective belief in "progress," the notion that, as Richard Slotkin puts it in his study of nineteenth-century American culture, "civilized power always takes the form of establishing human control over nature."[20] Faith in this power inspired C. G. H. Goss, president of the Indiana Engineering Society, when he exhorted his fellow engineers in 1895 to go forth and conquer nature and enslave it in order to "minister to the necessities and comforts of all mankind."[21] Belief in this same power animated Dixon Place's complacent certainty that draining the marsh would be a "great blessing to coming generations,"[22] and underlay William Whitten's pronouncement that draining was "almost a *necessity*" [emphasis in the original] because of the "prodigious development" which was the "destiny" of Chicago.[23] The logic of this power insists that human advancement can only come at the expense of nature, which, valueless in itself, must either be removed as an impediment, or exploited for human gain. The question is not whether values change—they do—but what values should replace the old ones?

As I pointed out in the introduction, the notion that nature has a value that transcends the Lockean idea of its "worthlessness" is relatively new. But it has wide currency. The 1982 United Nations World Charter for Nature begins, "Every form of life is unique, warranting respect regardless of its worth to man."[24] Many believe that natural value is not conferred by humans, but is a unique property held by all living things, that it is a condition of life. The philosopher Holmes Rolston III advocates for what he calls "objective natural value," which he believes is a product of biological conservation:

Biological conservation did not begin when the United Nations promulgated its World Charter for Nature, nor when the United States Congress passed the Endangered Species Act, nor even when Teddy Roosevelt withdrew forest reserves, nor even when Noah built the ark to save endangered species. Biological conservation began when life began, three and a half billion years ago. Those who do not conserve natural values are soon dead. Biological conservation in the deepest sense is not something that originates in the human mind, is modeled for Forplan programs on national forests, or written into acts of Congress. Biological conservation is innate as every organism conserves, values its life.[25]

Belief in natural value requires a re-examination of the relationship between humans and nature. This means an expansion of our ethical responsibilities to include nature, as well as fellow humans. Rolston argues persuasively that we need to form ethical judgments at the planetary level—the earth is home to the biotic community of life, after all, the *only* home for that community, and the only community that humans have.

We might begin the process of re-examination by thinking critically about Locke's claim that only human labor can impart value to nature, discussed in the Introduction. William Cronon puts that idea into perspective in commenting on the rapid growth of Chicago as market and broker to agricultural and logging interests in the nineteenth century:

> [T]he labor theory of value cannot by itself explain the astonishing accumulation of capital that accompanied Chicago's growth. Human labor may have been critical to planting, harvesting, and transporting the grain that passed through Chicago's elevators, or to logging, driving, and milling the lumber in its yards, but much of the value in such commodities came directly from first [original] nature, not second [processed], nature. The fertility of the prairie soils and the abundance of the northern forests had far less to do with human labor than with the autonomous ecological processes that people exploited on behalf of the human realm—a realm less of *production* than of *consumption* [emphasis in the original] ... The abundance that fueled Chicago's hinterland economy thus consisted largely of stored sunshine: this was the wealth of nature, and no human labor could create the value it contained. Although people might use it, redefine it, or even build a city from it, they did not produce it.[26]

Such an insight also puts into perspective the claim of many in the nineteenth century that the marsh was "useless," and had "injured" the land (see the Introduction). On the contrary, the marsh, not reclamation, using similar ecological processes to those Cronon refers to, created the conditions that made its soil some of the most productive farmland in the country.

Shifting the cultural locus of value from the human to the natural encounters formidable challenges from the ingrained notion that humans have an unquestioned right to exploit their own property as they please. Aldo Leopold understood this better than many environmentalists. Leopold was a professor of forestry and ecology whose lifelong advocacy for conservation was often directed at farmers—private landowners whose stewardship of their property, in Leopold's view, needed improvement. Leopold's conservationism, in other words, had a strong element of husbandry—the notion that owners of land should be watchful caretakers of the land's resources and its natural beauty. Leopold cautioned that government regulation and public ownership of land, though helpful to conservation, was not enough. He believed that individual landowners had the greatest responsibility:

> It is the individual farmer who must weave the greater part of the rug on which America stands. Shall we weave into it only the sober yarns which warm the feet, or also some of the colors which warm the eye and the heart? Granted there may be a question which returns him the most profit as an individual, can there be *any* question which is best for his community? [Emphasis in the original.] This raises the question: is the individual farmer capable of dedicating private land to uses which profit the community, even though they may not clearly profit him? We may be overhasty in assuming that he is not.[27]

Like Holmes Rolston, Leopold believed that humans are part of a biotic community, and that we have an ethical responsibility to the non-human members of that community. He defined his land ethic, often cited and discussed by environmentalists, not only in the context of the economics of land use, but also in the context of private ownership. Leopold believed that the only remedy for the idea that land had only economic value was for private owners to develop an ethical obligation to their land. He addresses them directly:

> [Q]uit thinking about decent land use as solely an economic problem. Examine each question in terms of what is ethically and aesthetically right, as well as what is economically expedient. A thing is right when it tends to preserve integrity, stability, and beauty of the biotic community. It is wrong when it tends otherwise.[28]

Leopold was not advocating for the dismissal of economics in determining land use. He noted the "fallacy" only of a certain *kind* of economic thinking:

> The fallacy the economic determinists have tied around our collective neck, and which we now need to cast off, is the belief that economics determine *all* land use

[emphasis in the original]. This is simply not true. An innumerable host of actions and attitudes, comprising perhaps the bulk of all land relations, is determined by the land users' tastes and predilections, rather than by his purse. The bulk of all land relations hinges on investments of time, forthought [sic], skill, and faith rather than on investments of cash. As a land-user thinketh, so is he.[29]

Leopold's ideas have much in common with Thomas Jefferson's view of the property owner's democratic and communitarian obligations, discussed in the Introduction. Leopold does not advocate for a greater role of government in the preservation of natural areas. Rather, he consistently stresses the responsibility of the individual landowner to the community of other landowners, and to the wider biotic community of soil, water, plants, and animals that we all share. He believed that such an ethic could only succeed as a product of social evolution, that its development must be the same as for any other ethic: "social approbation for right actions, social disapproval for wrong actions."[30] Achieving the realization of Leopold's land ethic though social evolution may be a long, difficult process. Competition from the economic and historical imperatives of private ownership is fierce. Cultural values long imbedded in individual attitudes toward property are not easily changed.

Individual values are intertwined with legal and political ones, making it especially difficult to protect vulnerable areas. Legal barriers are daunting. Arthur McEvoy points out that the "law of property in the United States contains a profound bias toward developmental uses and against such nonmarket values as the health and welfare of the communities that live on the land." He says that "the fundamental liberty of private owners to develop their property as they please is the cornerstone of American civil and economic freedom, while relatively unlimited access to the resources of the public lands is an all but inviolable principle in American politics."[31]

Conservationists need creative strategies to cope with these cultural and legal challenges. The US Department of Agriculture has found ways, using economic incentives, to move landowners in the direction of allowing portions of their property to be set aside for conservation. One of these programs, administered by the USDA's Natural Resources Conservation Service (NRCS), is specifically designed for wetlands conservation and has had success with Kankakee basin landowners. The Wetland Reserve Easement Program allows participants to retain ownership of their property, while the federal government maintains surface rights. This means that the property cannot be farmed, and must revert to undrained wetlands, which the NRCS can restore. There are thirty-year and permanent contracts available, though the permanent option is the more popular.[32] In the last twenty or thirty years, over a hundred easements, comprising nearly 13,000 acres, have been enrolled in the Kankakee River watershed, most of these along the Kankakee River.[33]

There is another hopeful sign that other things have changed along the Kankakee. Indiana farmers are not the only ones to enroll in this program; it is also designed to attract the participation of native tribes. In 2002, the NCRS helped the Potawatomis to return to the Kankakee Marsh when it joined with the Pokagon Band to restore over 1,400 acres of their ancestral lands.[34] Elan Pochedley points out that this restoration effort is one of several projects undertaken by Potowatomis and other tribes in Indiana and Michigan that are "emblematic of ecological initiatives that are led by and inclusive of Indigenous nations who retain obligations to the waters, lands, and nonhuman relatives of their traditional territories and homelands."[35] Native regard for the land shows that the fundamentals of Leopold's land ethic has a rich and potent ancestry.

These are hopeful signs of environmental recovery along the river, and *Everglades of the North* certainly raised public awareness about the Kankakee Marsh. It is an open question, however, whether or not attitudes toward conservation have changed among policymakers. In the nineteenth and twentieth centuries, agricultural interests were dominant in the move to destroy 97 percent of Indiana wetlands. In the twenty-first century, developers of residential and commercial property have taken over the role. In 2021 and again in 2024, the state government passed laws which allow further loss of wetlands for the development of subdivisions and commercial properties. These laws were designed specifically to loosen state restrictions on the draining of marshy areas. As in earlier times, economic interests were cited as the main reason for allowing the change in regulation. State Senator Rick Niemeyer, who sponsored the 2024 bill, explained that the permitting process was a burden for property owners because they "get pushed back for six months to a year before they can go forward and it costs them a lot of money."[36]

Economic interests still trump all other considerations for land use in the Indiana General Assembly. Regulation, so important to the preservation of natural areas, is created by legislatures, and can always be undone by legislatures. It can also be curtailed by the courts. In 2024, the US Supreme Court overturned a forty-year-old ruling allowing federal and state agencies broad power to implement regulatory law. This reversal at the federal level may have long-lasting, indeed, grave consequences to conservationist goals. We should remember that reaction to the regulatory policies of the 1960s and 1970s began early. Under President Ronald Reagan (1981–89), industries gained increased abilities to regulate themselves, and organizations such as the Environmental Protection Agency saw drastic budget cuts. Carolyn Merchant argues that "[a]nti-environmental forces" saw significant gains during the 1980s and 1990s.[37] Anti-wetland policies in Indiana, and the Supreme Court's recent decision, show that those forces are still very active, and very influential.

Wallace Stevens, famous poet, said not very famously that the purpose of poetry is to help us live our lives. Perhaps he meant that life is difficult, and poetry might help us cope. Stevens believed that living a successful life meant more than having economic prosperity. He knew that such a life was impossible without the cultivation of a sense of beauty. A commitment to Holmes Rolston's conservation of natural value emerges from the sense of natural beauty. After all, we are always inclined to protect and preserve that which we consider beautiful. Perhaps we can learn from many sources—the Potawatomis and nature itself—as well as the creators of beauty—poets and artists—the values we need to appreciate our ethical obligation to the earth.

## NOTES

1. "Region Streets Flooded," *The Times* (Munster, IN), February 21, 2018, and "Region Battles Flooding," *The Times* (Munster, IN), February 23, 2018. All newspaper articles, except as noted, were accessed online at Newspapers.com or Newspaperarchives.com.
2. Christopher B. Burke Engineering, *Kankakee River Flood and Sediment Management Work Plan* (Christopher B. Burke Engineering, August 2019), 23.
3. Dan Carden, "Lawmakers May Drain Kankakee River Board," *The Times* (Munster, IN), January 16, 2019.
4. Carden, "Lawmakers May Drain Kankakee River Board."
5. "Frequently Asked Questions," *River News: Journal of the Kankakee River Basin and Yellow Basin Development Commission*, no. 7 (August 2023), https://kankakeeandyellowrivers.org/news/newsletters/.
6. "Frequently Asked Questions," *River News*.
7. "Frequently Asked Questions," *River News*.
8. Burke Engineering, *Kankakee River Flood*, xi.
9. Burke Engineering, *Kankakee River Flood*, v.
10. Burke Engineering, *Kankakee River Flood*, 23.
11. See *River News: Journal of the Kankakee River Basin and Yellow Basin Development Commission*, no. 7 (August 2023), and no. 8 (December 1923): https://kankakeeandyellowrivers.org/news/newsletters/.
12. See *River News: Journal of the Kankakee River Basin*, nos. 7 and 8, https://kankakeeandyellowrivers.org/.
13. Conversation with Scott Pelath, executive director of the Kankakee River and Yellow Basin Development Commission, July 26, 2024.
14. Elizabeth Maree Hare, "Making Histories with Science: Paleoecology and Conservation in the Midwestern United States" (PhD diss., University of Santa Cruz, 2017), ProQuest (10262787), 95.
15. See *River News: Journal of the Kankakee River Basin and Yellow Basin Development Commission*, no. 7, August 2023, https://kankakeeandyellowrivers.org/.

16. See the Introduction for a discussion of Thomas Jefferson's ideas of democratic ownership.
17. William Whitten, "Kankakee Drainage," *Twelfth Annual Report of the Indiana Engineering Society, at its Twelfth Annual Meeting, Lafayette, Ind, Jan. 12, 13, and 14* (N.p., 1892), 84, 88. See chapter 7 for a full discussion of Campbell's changing views.
18. Hare, "Making Histories with Science," 100–101.
19. Hare, "Making Histories with Science," 101.
20. Richard Slotkin, *The Fatal Environment: The Myth of the Frontier in the Age of Industrialization, 1800–1890* (Atheneum, 1985), 214.
21. C. G. H. Goss, "Address by President Goss," *Proceedings of the Fifteenth Annual Meeting of the Indiana Engineering Society, Held at Indianapolis, IN, January 7, 9, and 10, 1895* (Indianapolis, 1895), 130.
22. Dixon Place, "Mr. Place Replies," *St. Joseph County Independent*, February 13, 1897.
23. Whitten, "Kankakee Drainage," 73.
24. Holmes Rolston, III, *Conserving Natural Value* (Columbia University Press, 1994), 167.
25. Rolston, III, *Conserving Natural Value*, 168.
26. William Cronon, *Nature's Metropolis: Chicago and the Great West* (Norton, 1991), 151. Cronon's reference to the labor theory of value here is to the economic ideas of Karl Marx and Adam Smith. However, John Locke provides the philosophical foundation for this idea.
27. Aldo Leopold, *The River of the Mother of God and Other Essays*, ed. J. Baird Callicott and Susan L. Flader (The University of Wisconsin Press, 1991), 260.
28. Leopold, *A Sand County Almanac and Sketches Here and There* (Oxford University Press, 1949), 224.
29. Leopold, *A Sand County Almanac*, 225.
30. Leopold, *A Sand County Almanac*, 225.
31. Arthus McEvoy, "Markets and Ethics in U.S. Property Law," in *Who Owns America? Social Conflict Over Property Rights*, ed. Harvey M. Jacobs (University of Wisconsin Press, 2000), ebook edition, 75.
32. Natural Resources Conservation Service, US Department of Agriculture, "Wetland Reserve Easements," accessed August 20, 2024, https://www.nrcs.usda.gov/programs-initiatives/wre-wetland-reserve-easements.
33. From an anonymous USDA source.
34. Brandon O'Connor, "Fridays on the Farm: Restoring the Grand Kankakee Marsh," accessed August 20, 2024, https://www.farmers.gov/blog/fridays-on-farm-restoring-grand-kankakee-marsh.
35. Elan Pochedley, "Restorative Cartography of the Theakiki Region: Mapping Potawatomi Presences in Indiana," *Open Rivers: Rethinking Water, Place and Community*, no. 18 (Spring 2021): 39.

36. Casey Smith, "Bill Further Rolling Back Indiana Wetland Protection is First to Land on Governor's Desk," *Indiana Capital Chronicle*, February 7, 2024, indianacapitalchronicle.com.
37. Carolyn Merchant, *The Columbia Guide to American Environmental History* (Columbia University Press, 2002), 182–83.

# BIBLIOGRAPHY

Following is a list of books, government documents, engineering reports, and journal articles consulted for this book. References to the many newspapers cited can be found in the notes to each chapter. All newspaper articles, except as noted, were accessed online at Newspapers.com or Newspaperarchives.com.

Ade, John. *Newton County, 1853–1911*. Bobbs-Merrill, 1911.
"Annual Report of the Auditor of State for the State of Indiana for the year 1854." In *Documents of the General Assembly of Indiana at the Thirty-Eighth Session*. Indianapolis, 1855.
"Annual Report of the Auditor of State of Indiana to the Legislature 1860." In *Documents of the General Assembly of Indiana at the Forty-First Session, January 10, 1861*. Indianapolis, 1861.
*Annual Report of the Board of Commissioners for the Removal of the Limestone Ledge in the Kankakee River, 1890*. Indianapolis, 1890.
*Annual Report of the Board of Commissioners to Remove the Rock Ledge in the Kankakee, 1893*. Indianapolis, 1893.
Barce, Elmore. *Beaver Lake, A Land of Enchantment*. The Kentland Democrat, 1938.
Bartlett, Charles H. *Tales of Kankakee Land*. Scribner, 1907.
Battey, F. A. *Counties of Warren, Benton, Jasper and Newton, Indiana*. F. A. Battey, 1883.
Bhowmik, Nani G., et al. "Hydraulics of Flow and Sediment Transport in the Kankakee River in Illinois." In *Report of Investigation 98, State of Illinois, State Water Survey Division*. N.p., 1980.
*Brevier Legislative Reports, Volume V 1861*. Indianapolis, 1861. https://purl.dlib.indiana.edu/iudl/law/brevier/VAA8558-05.
*Brevier Legislative Reports, Volume XI, 1869*. Indianapolis, 1869. https://purl.dlib.indiana.edu/iudl/law/brevier/VAA8558-11.
*Brevier Legislative Reports, Volume XII, 1871*. Indianapolis, 1871. https://purl.dlib.indiana.edu/iudl/law/brevier/VAA8558-12.
*Brevier Legislative Reports, Volume XIII, 1872*. Indianapolis, 1872. https://purl.dlib.indiana.edu/iudl/law/brevier/VAA8558-13.
Bright, Jesse D. "Some Letters of Jesse D. Bright to William H. English: 1842–1863." *Indiana Magazine of History* 30, no. 3 (September 1, 1934): 371–92.
Brookes, Andrew. *Channelized Rivers: Perspectives for Environmental Management*. John Wiley and Sons, 1988.
Burke Engineering, LLC, Christopher B. *Kankakee River Flood and Sediment Manage-*

*ment Work Plan*. Christopher B. Burke Engineering, LLC, 2019.

Burroughs, Burt. *Tales of an "Old Border Town" and along the Kankakee*. Benton Review Shop, 1925.

Campbell, John L. "How to Use the Waters of the Kankakee." In *Fourteenth Annual Meeting of the Indiana Engineering Society held at Indianapolis, Indiana, January 2, 3 and 4, 1894*. South Bend, IN, 1895.

Campbell, John L. *Report Upon the Improvement of the Kankakee River and the Drainage of the Marsh Lands of Indiana*. Indianapolis, 1883.

Campion, Thomas. "Indian Removal and the Transformation of Northern Indiana." *Indiana Magazine of History* 107, no. 1 (March 2011): 32–62.

"Civil Works Administration." *National Park Service*. Accessed January 7, 2024. https://www.nps.gov/articles/000/civil-works-administration.htm.

Cronon, William. *Changes in the Land: Indians, Colonists, and the Ecology of New England*. Hill and Wang, 2003.

Cronon, William. *Nature's Metropolis: Chicago and the Great West*. Norton, 1991.

Daniels, Rev. E. D. *A Twentieth Century History and Biographical Record of LaPorte County, Indiana*. Lewis Publishing, 1904.

Doyle, Martin. *The Source: How Rivers Made America and America Remade its Rivers*. W. W. Norton, 2018.

*Fourth Annual Report of the Department of Public Works and Buildings, Division of Waterways, July 1, 1920 to June 30, 1921*. N.p., 1921.

Gates, P. W. "Hoosier Cattle Kings." *Indiana Magazine of History* 44, no. 1 (March 1948): 1–24.

Gordon, Leon M., II. "The Price of Isolation in Northern Indiana, 1830–1860." *Indiana Magazine of History* 46, no. 2 (June 1950): 39–60.

Goss, C. G. H. "Address by President Goss." In *Proceedings of the Fifteenth Annual Meeting of the Indiana Engineering Society, Held at Indianapolis, IN, January 7, 9, and 10, 1895*. Indianapolis, 1895.

"Grand Kankakee Marsh National Wildlife Refuge: Preliminary Project Proposal Summary." US Fish and Wildlife Service, 1996.

Greenberg, Joel. *A Natural History of the Chicago Region*. University of Chicago Press, 2002.

Greenberg, Joel, ed. *Of Prairies, Woods, & Water: Two Centuries of Chicago Nature Writing*. University of Chicago Press, 2008.

Gwynne, S. C. *Empire of the Summer Moon: Quanah Parker and the Rise and Fall of the Comanches*. Scribner, 2010.

Hamilton, Louis H., and William Darroch, eds. *A Standard History of Jasper and Newton Counties, Indiana*. Lewis Publishing, 1916.

Hare, Elizabeth Maree. "Making Histories with Science: Paleoecology and Conservation in the Midwestern United States." PhD diss., University of Santa Cruz, 2017, ProQuest (10262787).

Heasley, Lynn. *A Thousand Pieces of Paradise: Landscape and Property in the Kickapoo Valley*. University of Wisconsin Press, 2005.

*History of LaPorte County, Indiana*. Chicago, 1880.

*History of Porter County, Indiana*. Vol. 1. Lewis Publishing, 1912.

Hoffman, Len. Foreword to "The Old Kankakee: The Dream of 100,000 Hoosiers Can Come True." Indiana Division of the Izaak Walton League, 1942.

Howard, Timothy. *A History of St. Joseph County, Indiana*. Vol. 2. Lewis Publishing, 1907.

Howat, William Frederick. *A Standard History of Lake County, Indiana and the Calumet Region*. Vol. 1. Lewis Publishing, 1915.

*Indiana Code, Title 18, City and Town Government, 1977*. Indiana Code Historical Statutes. https://iuidigital.contentdm.oclc.org/digital/collection/IC/id/88963/rec/17.

Indiana Department of Conservation. "The Kankakee Basin Plan." In "The Old Kankakee: The Dream of 100,000 Hoosiers Can Come True." Izaak Walton League, 1942.

Ivens, J. Loreena, Nani G. Bhowmik, Allison R. Brigham, and David L. Gross. *The Kankakee River Yesterday and Today*. Illinois Department of Energy and Natural Resources, 1981.

*Izaak Walton League of America: A Century of Conservation Leadership, 1922–2022*. Izaak Walton League of America, 2022.

Jakle, John A. "Toward a Geographical History of Indiana: Landscape and Place in the Historical Imagination." *Indiana Magazine of History* 89, no. 3 (September 1993): 1–33.

*Journal of the Indiana State Senate, 40th Session, January 6, 1859*. Indianapolis, 1859.

Judson, W. V. *Report of the Board of Engineers for Rivers and Harbors*. House of Representatives Document 931, 64th Congress, 1st Session, March 20, 1916.

*Laws of the State of Illinois Enacted by the Fifty-Second General Assembly at the Regular Biennial Session*. Schnept and Barnes, 1921.

*Laws of the State of Indiana, Fifty-Second Regular Session, 1881*. Indianapolis, 1881.

*Laws of the State of Indiana, Forty-Ninth Session, 1875*. Indianapolis, 1875.

*Laws of the State of Indiana, Forty-Seventh Regular Session*. Indianapolis, 1871.

*Laws of the State of Indiana, Forty-Sixth Regular Session of the General Assembly, 1869*. Indianapolis, 1869.

*Laws of the State of Indiana, Passed and Published, at the Regular Session of the 58th Regular Session of the General Assembly*. Indianapolis, 1893.

*Laws of the State of Indiana, Passed at the Seventy-First Regular Session of the General Assembly, 1919*. Indianapolis, 1919.

Leopold, Aldo. *The River of the Mother of God and Other Essays*. Edited by J. Baird Callicott and Susan L. Flader. University of Wisconsin Press, 1991.

Leopold, Aldo. *A Sand County Almanac and Sketches Here and There*. Oxford University Press, 1949.

Locke, John. *John Locke, An Essay Concerning Human Understanding*. Vol. 1. Collated by Alexander Fraser. Oxford, 1894.

Locke, John. *John Locke, Two Treatises of Government*. Prepared by Rod Hay for the McMaster University Archive of the History of Economic Thought. London, 1823.

May, Jill, and Robert May. *Spearheading Environmental Change: The Legacy of Indiana Congressman Floyd J. Fithian*. Purdue University Press, 2022.

McCool, Daniel. *River Republic: The Fall and Rise of America's Rivers*. Columbia University Press, 2012.

McCorvie, Mary R,. and Christopher L. Lant. "Drainage District Formation and the Loss of American Wetlands, 1850–1930." *Agricultural History* 67, no. 4 (Autumn 1993): 13–39.

McEvoy, Arthur. "Markets and Ethics in U.S. Property Law." In *Who Owns America? Social Conflict Over Property Rights*, edited by Harvey M. Jacobs. University of Wisconsin Press, 2000.

Mckee, Irving. "The Centennial of the Trail of Death." *Indiana Magazine of History* 35, no. 1 (March 1939): 7–16.

Merchant, Carolyn. *The Columbia Guide to American Environmental History*. Columbia University Press, 2002.

Meyer, Alfred H. "The Kankakee 'Marsh' of Northern Indiana and Illinois." PhD diss., University of Michigan, 1936. Papers of the Michigan Academy of Science, Arts and Letters, vol. 21.

Milk, Lemuel. "The Petition of Lemuel Milk, Jane A. Milk and Henry H. Cooley to the General Assembly of the State of Indiana, in Reference to Quieting Title to Beaver Lake Land." Indianapolis, 1881.

"Minutes of the Board of Commissioners for the Removal of the Limestone Ledge in the Kankakee River." Indiana State Archives, July 19, 1889.

Nichols, Fay Folsom. *The Kankakee: Chronicle of an Indiana River and its Fabled Marshes*. Theo. Gaus' Sons, 1965.

Pence, W. D., and Morton Downey. *A Report Upon the Drainage of Agricultural Lands in the Kankakee Valley, Indiana, U.S. Department of Agriculture Circular 80*. Government Printing Office, 1909.

Perry, J. L., Chief Engineer. "Progress Report: Investigation of Kankakee and Yellow Rivers, Indiana for Flood Control and Major Drainage." Indiana Flood Control and Water Resources Commission, January 1957.

Pochedley, Elan. "Restorative Cartography of the Theakiki Region: Mapping Potawatomi Presences in Indiana." *Open Rivers: Rethinking Water, Place and Community*, no. 18 (Spring 2021): 1–49.

Prince, Hugh. *Wetlands of the American Midwest, A Historical Geography of Changing Attitudes*. University of Chicago Press, 1997.

*Prospectus and Articles of Association of the Kankakee Valley Draining Co. of Indiana*. New York, 1869.

*Prospectus of the Kankakee Valley Draining Co. of Indiana*. LaPorte, IN, 1871.

Reed, Earl H. *Tales of a Vanishing River*. John Lane, 1920.

Reiger, John F. *American Sportsmen and the Origins of Conservation*. 3rd ed. Revised and expanded. Oregon State University Press, 2001.

"Report and Testimony of the Swamp Land Committee, 1863." In *Documents of the General Assembly of Indiana at the 42nd Regular Session, January 8, 1863, Part 2, Vol II*. Indianapolis, 1863.

"Report from the Chief of Engineers on the Kankakee River, ILL and IND, Covering Navigation, Flood Control, Power Development, and Irrigation." In *Report to the House of Representatives, Document 784, February 26, 1931*. Kankakee River Basin Commission Archives, Kankakee River Basin and Yellow River Development Commission, Valparaiso, IN.

"Report [to the Senate] of the Commissioner of the General Land Office Relative to the Drainage of Beaver Lake, Indiana [1872]." In *The Executive Documents for The Senate of the United States for the Second Session of the Forty-Second Congress, 1871–'72*. Government printing office, 1872.

*Report of the Board of Commissioners for the Removal of the Limestone Ledge in the Kankakee River*. Indianapolis, 1895.

*Report of the Swamp Land Committee to the General Assembly of the State of Indiana, 1859*. Indianapolis, 1859.

*Report on the Water and Related Land Resources, Kankakee River Basin, Indiana, State of Indiana*. US Department of Agriculture, US Department of the Interior, November 1976.

*Reports of Cases Argued and Determined in the Supreme Court of Judicature of the State of Indiana, vol. XXXI*. Indianapolis, 1883.

"Review Summary of the Report to the Governor." In *Kankakee River Task Force Report to Illinois Governor, 1978*. From the Kankakee River Basin Commission Archives, Kankakee River Basin and Yellow River Development Commission, Valparaiso, IN.

Rolston, Holmes, III. *Conserving Natural Value*. Columbia University Press, 1994.

Roosevelt, Theodore. "Publicizing Conservation at the White House." In *American Environmentalism: Readings in Conservation History*, edited by Roderick Frazier

Nash. 3rd ed. McGraw Hill, 1990.

Schweitzer, Art, Dave Andrews, and Rich and Betty Jonas. *Schererville Through the Years: A Pictorial Look Back*. Schererville Historical Society, 2002.

SEG Engineers and Consultants. *Kankakee River Master Plan: A Guide for Flood Control and Land Use Alternatives in Indiana*. SEG Engineers and Consultants, 1989.

Sherland, Marilyn Hiatt, ed. *The Island: Land Between the Kankakee and Pine Creek*. Walkerton Area Historical Society, 2007.

Slotkin, Richard. *The Fatal Environment: The Myth of the Frontier in the Age of Industrialization, 1800–1890*. Atheneum, 1985.

Smithsonian. National Museum of African American History and Culture. "The Great Mississippi River Flood of 1927." Accessed January 7, 2024. https://nmaahc.si.edu/explore/stories/great-mississippi-river-flood-1927.

"Status of Investigation for Flood Control and Major Drainage: Kankakee and Yellow Rivers, Indiana." Indiana Flood Control and Water Resources Commission, 1960.

Strausberg, Stephen. "Indiana and the Swamp Land Act: A Study in State Administration." *Indiana Magazine of History* 73, no. 3 (September 1977): 191–203.

Surface-Evans, Sarah. "Intra-Wetland Land Use in The Kankakee Marsh Region of Northwestern Indiana." *Midcontinental Journal of Archeology* 40, no. 2 (Summer, 2015): 166–89.

Urban, Michael. "An Uninhabited Waste: Transforming the Grand Prairie in Nineteenth Century Illinois." *Journal of Historical Geography* 31 (2005): 647–65.

Voight, William, Jr. *Born with Fists Doubled: Defending Outdoor America*. Izaak Walton League Endowment, 1992.

Werich, J. Lorenzo. *Pioneer Hunters of the Kankakee*. N.p., 1920.

"Wetland Reserve Easements." Natural Resources Conservation Service, US Department of Agriculture. Accessed August 20, 2024. https://www.nrcs.usda.gov/programs-initiatives/wre-wetland-reserve-easements.

Whitaker, John, and Charles Amlaner, Jr., eds. *Habitats and Ecological Communities of Indiana, Presettlement to Present*. Indiana University Press, 2012.

Whitten, William. "Report of the Committee on Drainage." In *Proceedings of the Thirteenth Annual Meeting of the Indiana Engineering Society Held at South Bend, Indiana, January 23, 24, & 25, 1893*. South Bend, IN, 1893.

Whitten, William. "Kankakee Drainage." *Twelfth Annual Report to the Indiana Society of Engineers, Held at Lafayette, Ind., Jan. 12, 13, and 14, 1892*. N.p., 1892.

Whitten, William. "Report from President Whitten." In *Proceedings of the Fourteenth Annual Meeting of the Indiana Engineering Society held at Indianapolis, Indiana, January 2, 3 and 4, 1894*. South Bend, IN, 1895.

# INDEX

Ade, John, 18
agriculture, 91, 121–24, 145–47, 204; drainage supporting, 22–23, 93, 95, 116, 179; environment over, 184; land for, 2, 124, 181; runoff from, 10n21; US Department of Agriculture, 144, 206. *See also* farmland
algae blooms, 10n21
*American Sportsmen and the Origin of Conservation* (Reiger), 118
*anthropogeomorphology*, 112
assessments, for land, 55–56, 57, 58, 59–60
Auditor's Report of 1860, 35, 48n47
Aylesworth, Michael, 168, 169–70, 173, 178, 182, 185–86

Barce, Elmore, 19
Barker, Ned, 19–20
Bartlett, Charles H., 16–17
Beal, Frank, 152–53, 155
Beaver Lake, 1, 3, 13, 201, 202, 203; Ade on, 18; Campbell on, 64, 65; destruction of, 33, 51, 201; drainage of, 23, 28, 51; landowners relationship with, 8; lands, 28, 36–37, 39–44, 47n33, 47n37, 51; pre-emptors of, 37–38, 39, 40; as public domain, 23, 29, 36, 44; sign memorializing, 27, 28; surveyors of, 28–29, 45n8; Test on, 20; US Congress and, 39, 42; waterfowl on, 14, 19–20, 44, 117; wildlife in, 14. *See also* Bright, Michael G.
*Beaver Lake* (Barce), 19
Beaver Lake Basin, 45, 125, 188
Beaver Lake Ditch, 32–33, 34, 36, 42, 45
Benda, George, 147–49
Bhowmik, Nani, 170, 171

biological conservation, 203–4
Blake, William, 35
Board of Aldermen, Momence, Illinois, 81
Board of Commissioners for the Removal of the Limestone Ledge in the Kankakee River, 72–73, 76, 78, 79, 89; Whitten reporting to, 81–82, 85n42
Board of Engineers for Rivers and Harbors, US, 99
Boissy, Fred, 137, 138, 143, 148, 151, 163
Bradley, James, 57–58
*Brazil Daily Times* (newspaper), 115
Brigham, Allison, 170–71
Bright, George, 35
Bright, Jesse D., 33
Bright, Michael G., 33–34, 36–38, 47n34, 47n37, 47n40; Auditor's Report of 1860 and, 35, 48n47; phony title of, 39–42, 43–44, 45n8, 48n61, 49n68, 51, 78
Brown, John, 76, 78, 101
Brown, Lytle, 134
built environment, 28, 45, 112–13, 134, 201
Bunyan, John, 21
Burke Report, 198–200
Burroughs, Burt, 19, 117
Buyer, Steve, 188–90, 191–92

CAFO. *See* confined animal feeding operation
Campbell, John L., 75–76, 200; general plan advocated for by, 87, 129, 130–32; proposal for straightening of Kankakee River, 64–66, 88–89, 92, 129; *Report on the Improvement of the Kankakee River*, 73, 139n6; writing to Indiana Engineering Society, 82–83

218 / INDEX

Cass, Charles, 71–72, 76; Kankakee County Court and, 74, 75; rock ledge removal and, 79–80
Cass, George W., 74, 103; Kankakee Valley Draining Company and, 60–61, 71; on natural dam, 62
Cass, Lewis, 60, 71
Cass, Mary, 71–72, 76; Kankakee County Court and, 74, 75; rock ledge removal and, 79–80
channelization, of Kanakee River, 101, 112, 130–31; Campbell on, 200; Doyle on, 132–33; ecological damage from, 133, 150–51, 181; to state line, 115
Chesak, Allen, 184, 187
Chicago, Illinois, 90–91, 111, 136
Chicago and Eastern Illinois Railroad, 71, 74–75, 77, 80
*Chicago Daily Tribune* (newspaper), 37; Kruse interviewed, 190–91; on removal of rock ledge, 80, 89
chorography map, 112–13
Citizens for the Preservation of the Kankakee Basin, 184–85, 187
Civilian Conservation Corps, 120, 122, 164
Civil Works Administration, 120, 121, 122
Clean Water Act of 1972, 133, 164–65
clearing, of Kankakee River: Dovas on, 177–78; project for, 163–64, 165–66, 168–69, 171–73, 200; snagging and, 152, 153, 154, 159, 160–62
Coffin, W. H. H., 96, 102
Collins, William F., 121–23, 124
Committee on Public Lands, in House of Representatives, 41
Condit, Amzie B., 32–33, 36, 38–40, 47n33, 47n34, 51
confined animal feeding operation (CAFO), 45

*Congressional Globe* (1850), 22
Conner, J. W., 38–39
consequences, of drainage, 1, 161, 181, 200, 207; environmental, 23, 65, 168
conservation: Civilian Conservation Corps, 120, 122, 164; fish and game, 115; Governors' Conference on Conservation, 116, 117; Indiana Department of Conservation, 115, 116, 120, 123–25; Leopold on, 205, 206; money for, 116, 178, 187; Roosevelt, T., and, 116, 117, 118; SCS, 133, 137, 147, 177, 181, 198; soil, 152, 154; US Army Corps of Engineers and, 164–65; values of, 3
Conservation Act (1919), 115–16
Conservation Defense League, 160
Cook, Edward, 185
Cook, W. F., 96
Cooley, Henry, 36, 42, 77–78, 80
Cronon, William, 10n22, 90–91, 111, 204, 209n26
cropland, 23, 44, 55, 144, 146; Campbell encouraging, 65; flooding of, 136–37, 149–50, 169
*Crown Point Register* (newspaper), 59–60
Crumpacker, Senator, 88
current, of Kankakee River, 2, 52, 60–61, 72–73, 89, 131

*Daily Journal* (newspaper), Illinois, 162, 171
*Daily Tribune* (newspaper), 77
dams: Chicago and Eastern Illinois Railroad and, 74; natural, 14, 60–61, 62, 65, 71
Darroch, John, 30, 32, 39
Daube, Michael, 187
Davos, Christos, 172
Dean, Algy, 36, 37, 57
Declaration of Independence, 4

Des Plaines river, 2
destruction: of Beaver Lake, 33, 51, 201; of Kankakee Marsh, 1, 55, 201
Dickens, Charles, 21–22
dikes, 134, 135, 136–37, 149, 159–60, 169; breaking, 183; repair of, 161, 165
Dilg, Will H., 119–20
ditches: Beaver Lake Ditch, 32–33, 34, 36, 42, 45; Elsbree Ditch, 99; Lake County, 90; lateral, 55, 65–66, 91–92, 95, 96, 100, 129, 186; Marble–Powers Ditch, 101–2, 138; Miller Ditch, 97, 98; money for, 87, 96–97; open ditching, 30–31, 51; Place Ditch, 92, 95–96, 98, 100–102, 103, 106n38; Williams Ditch, 102, 103–4
Division of Waterways, 105, 108n81
divisions, between Illinois and Indiana, 99, 105, 154, 200
DNR. *See* Indiana Department of Natural Resources
Dodd, John, 32
Dovas, Christos, 177–78
Doyle, Martin, 132–33
drainage: agriculture supported by, 22–23, 93, 95, 116, 179; of Beaver Lake, 23, 28, 51; economic value of, 64–65; fish population reduced from, 113, 114; Illinois Drainage Commission, 77, 80; Izaak Walton League and, 118–20; of Kankakee Marsh, 1, 2, 51–53, 72; Marble in support of, 100–101; money for, 44, 53–54, 57, 66, 72, 88; Place on, 91–94; projects for, 2, 23, 88, 89–90; Rinehart on, 94; Whitten report on, 87–88. *See also* Kankakee Valley Draining Company; rock ledge, in Momence, Illinois
Drainage Act of 1869, 55–57, 88
drainage law, of 1869, 55–56, 88; Johnson disputing, 62–63; opposition meetings for, 58–60; repeal of, 63–64
drainage legislation of 1875 and 1881, 84
Drains and Levies Corporation Act of 1852, 53, 54, 88
dredging, of Kankakee River, 53, 59, 60, 63, 92, 105; cooperative, 99; Kankakee Reclamation Company, 96, 97–99, 106n34; Kankakee River Improvement Company, 94–96, 98; McWilliams dredging company, 101; opposition to, 103–4, 184. *See also* steam dredge boats
Dresser, Isaac Marker, 36
Dresser, Parker, 39
Dunn, John, 33, 35, 38–39, 41, 47n34, 49n68; on Board of Commissioners for the Removal of the Limestone Ledge in the Kankakee River, 73; resignation of, 76
Dustin, Tom, 160

Eagle Point island, 17
Earth Day, 165
economic value: of drainage, 64–65; of land use, 205–6, 207; motivations towards, 5; over preservation of environment, 2–3, 8, 119
ecosystems, of Kankakee Marsh, 1, 111
Eisenhower, Dwight D., 136
Elsbree Ditch, 99
Engels, Bob, 120–21, 122
environment: built, 28, 45, 112–13, 134, 201; channelization as causing damage to, 133, 150–51, 181; consequences to, 23, 65, 168; damage to, 132, 164, 173; economic value over, 2–3, 8, 119; environmentalists for, 145, 160, 161–62, 164–65, 172, 185; over agriculture, 184. *See also* erosion; flooding

environmentalists, 145, 160, 164–65, 172; KRBC targeted by, 161–62; wide-levee projects supported by, 185

environmental legislation: Clean Water Act of 1972, 133, 164–65; Conservation Act, 115–16; Drainage Act of 1869, 55–57, 88; Drains and Levies Corporation Act of 1852, 53, 54, 88; Federal Water Pollution Act of 1972, 165; National Environmental Policy Act of 1969, 165, 189; Swamp Land Acts of 1850, 22–23, 30, 54, 65, 88, 93

Environmental Protection Agency, 165, 166, 207

erosion, 1, 98, 117, 124, 147–48, 155, 171; flooding and, 172, 177, 179, 198; sediment and, 144, 145, 170, 199

*An Essay Concerning Human Understanding* (Locke), 4

*Everglades of the North* (documentary), 202–3, 207

farmland, 1, 6, 171, 187; flooding of, 134, 167–68, 197; Indiana Farm Bureau, 191; in LaPorte County, 97; productive, 3, 31, 32, 90, 115, 184, 204; United Nations and, 190–91; wetland turned to, 23; wide-levee project and, 184

Federal Water Pollution Act of 1972, 165

Final Environmental Statement, of US Army Corps of Engineers, 172

Fish and Wildlife Division, DNR, 162–63

fishing, 168–69; drainage and population for, 113, 114; Indiana Commissioner of Fish and Game, 113–14; US Fish and Wildlife Service, 122, 166, 187–88, 189–90

Fithian, Floyd, 137–38, 145–46

flood channel, 132

Flood Control Act of 1936, 133

flood control projects, KRBC, 144, 147–48, 171–72, 180, 188, 197–98

flooding, 1, 182–83; Burke Report on, 198–200; Campbell warning of, 130, 132; control for, 124, 134–35, 138, 144–48, 150–52, 161–63, 167–68; of cropland, 136–37, 149–50, 169; erosion and, 172, 177, 179, 198; of farmland, 134, 167–68, 197; of Kankakee Marsh, 116, 133–36; *Kankakee River and Sediment Work Plan* and, 198–99; management for, 180, 197, 198; *Program for the Future*, 178–80, 186; runoff, 179, 182; Whitten on, 151; Yellow River, 136, 149, 159–60. *See* channelization, of Kankakee River; levees

flood plain zoning ordinances, 145

flood protection systems, 145–46

flow velocity, 132

Frasier, George, 54, 57

fraudulent ownership, 31–32, 38; phony title and, 39–42, 43–44, 45n8, 48n61, 49n68, 51, 78

"Fundament of the Kankakee Marsh," 112

General Assembly of Indiana, 51–52, 159–60, 161, 192, 197, 207; Indiana Flood Control and Water Resources Commission and, 136; Jasper County investigated by, 32; KRBC and, 138–39, 185–86; open ditching system used by, 31; on restoration for Kankakee Marsh, 124–25; on straightening of Kankakee River, 104; wetlands and, 30

Gillman, Cletus, 137–38

glacial drift deposits, 104

Glidden, J. G., 60

Goodnow, Samuel, 29
*Goshen Weekly News* (newspaper), 89
Goss, C. G. H., 1, 4, 203
Governors' Conference on Conservation (1908), 116, 117
Grand Kankakee National Wildlife Refuge, 188
Graves, Gordon, 151, 153–54, 171
Gray, Isaac, 66, 87
Great Dismal Swamp, 17
Great Flood (1927), of Mississippi River, 121
Greenberg, Joel, 14, 15, 123
Gross, David, 171
Guard, Ray, 160
Gutwein, Steve, 197

*Hagerstown Exponent* (newspaper), 89
Hannah, William C., 55–56, 57, 60, 61–62
Hare, Elizabeth Maree, 201, 202
Hart, Aaron, 6
Heasley, Lynn, 7–8
Hendricks, 39
Hodges, Harold, 184–85
Hofmann, Len, 123–24
*Hope Pioneer* (newspaper), 71
Hough, Emerson, 119, 120, 122
House of Representatives, 39, 40, 41, 56, 62, 63, 108n69
Hudak, Dave, 190
Hunter, Judge, 19, 117
hunters: hunting clubs established for, 21; on Kankakee Marsh, 15–16, 19–20; *Pioneer Hunters of the Kankakee*, 13; steam dredge boats impacting, 118
hunting clubs, 15, 21
Husar, John, 190–91
hydrology, 51, 131, 132, 133, 172, 181

Illinois: attorney general of, 77; Chicago and Eastern Illinois Railroad, 71, 74–75, 77, 80; Drainage Commission, 77, 80; Kankakee County in, 2, 71; Momence, 2; State Budget Agency, 161; US Army Corps of Engineers sued by, 165–66, 167. *See also* Chicago, Illinois; Kankakee Marsh; Kankakee River
Illinois Department of Natural Resources, 2
Illinois Institute of Natural Resources, 152
Illinois River, 2, 77–78
Illinois Second District Appellate Court, 74, 79
Indiana: attorney general of, 77; Beaver Lake in, 1, 3; Jasper County, 29, 30, 32, 35, 47n37, 101, 151, 169; Kankakee Marsh in, 1, 2; Lake County, 18, 27, 90, 135, 137, 159, 179,183; landowners, 1, 3; LaPorte County, 18, 58–60, 92, 95, 97–98, 183; Newton County, 13, 18–19, 27, 42; Porter County, 18; Potawatomi tribe in, 5, 7, 13, 207; *The Report on the Water and Related Land Resources of the Kankakee River Basin*, 144–45, 146; rock ledge and, 72–73; Shelby, 134, 135, 136, 149, 159; South Bend, 13, 14, 183; Starke County, 17, 92, 96–97, 98–100, 101, 149, 190; State Budget Committee for, 183; state legislature of, 35, 118, 139; St. Joseph County, 17, 54, 55, 57, 58, 60, 97–98; Valparaiso, 59, 66, 145, 148, 182, 184–85. *See also* General Assembly of Indiana
Indiana Commissioner of Fish and Game, 113–14
Indiana Conservation Council, 160

222 / INDEX

Indiana Department of Conservation, 115, 116, 120, 123–25

Indiana Department of Natural Resources (DNR), 125, 137, 138, 144, 160, 187; endorsement of wide-levee project, 180–81, 182, 183, 184; Fish and Wildlife Division of, 162–63; KRBC and, 161, 177, 178, 180; on *Program for the Future*, 178–80, 186; US Army Corps of Engineers and, 162, 178, 180; West and, 146–47

Indiana Engineering Society, 129, 200, 203; Campbell writing to, 82–83; Goss addressing, 1; Whitten reporting to, 76, 82

Indiana Farm Bureau, 191

Indiana Flood Control and Water Resources Commission, 136

Indiana Grand Marsh Restoration Project, 188

*Indianapolis Daily Journal* (newspaper), 52, 53, 55

*Indianapolis Journal* (newspaper), 33, 37–38, 60, 62, 103

*Indianapolis News* (newspaper), 81, 100, 191

*Indianapolis Sentinel* (newspaper), 8, 9n4

Indiana Regional Planning Commission (NIRPC), 138

*Indiana State Sentinel* (newspaper), 52, 53

Indian removal, 20–21, 71

inert matter, 8, 9

Iroquois River, 170, 192, 198, 199

Izaak Walton League, 114, 115, 118–21, 122–23, 125; John Birch Society and, 138; KRBC and, 160, 167

Jakle, John, 28

Jasper County, Indiana, 29, 30, 35, 47n37, 101, 169; Boissy defending, 151; General Assembly investigating, 32

Jasper–Pulaski State Game Reservation, 116

Jefferson, Thomas, 4, 7, 8, 164, 206

J. J. Queally Company, 59

John Birch Society, 138

Johnson, Edwin T., 62–63

Jones, Aquilla, 35

Jontz, Jim, 185

*The Kankakee* (Nichols), 13

"The Kankakee Basin Plan," 123

Kankakee County, Illinois, 2, 71

Kankakee County Court, 74, 75, 77

Kankakee–Iroquois Regional Planning Commission, 191

Kankakee Marsh: Bartlett on, 16–17; Campbell on, 64–65; destruction of, 1, 55, 201; drainage of, 1, 2, 51–53, 72; Eagle Point island on, 17; ecosystems of, 1, 111; flooding of, 116, 133–36; General Assembly of Indiana on, 124–25; history of, 201–3; Ling on, 16; Meyer on, 14–16, 21; Meyers maps of, 112–13, 114; *Pioneer Hunters of the Kankakee* on, 13; railroads crossing, 91; restoration areas for, 125; *Tales of a Vanishing River* on, 13–14; topographical features of, 14–15, 21. See also ecosystems

"The Kankakee 'Marsh' of Northern Illinois and Indiana" (Meyer), 111–12

Kankakee National Wildlife Refuge, 188, 192, 203

Kankakee Reclamation Company, 96, 97–99, 106n34

Kankakee River, 1, 13; algae blooms in, 10n21; clearing and snagging of, 152,

153, 154, 159, 160–62; current of, 2, 52, 60–61, 72–73, 89, 131; damming of, 2; Greenberg on, 14; landowners relationship with, 8; main channel of, 65, 82, 92, 146, 153, 168, 179; management of, 179, 182; navigability of, 2, 165, 167; political forces and, 172–73; project for clearing and snagging of, 163–64, 165–66, 168–69, 171–73, 200; rock ledge blocking, 71–72; sedimentation, 148, 152–53, 170–71, 199; *Tales of an "Old Border Town" and along the Kankakee*, 19, 117; Walker and, 51–54. *See also* dredging, of Kankakee River; straightening, of Kankakee River

*Kankakee River and Sediment Work Plan*, 198–99

Kankakee River Basin and Yellow River Basin Development Commission, 198–99

Kankakee River Basin Commission (KRBC), 139, 146, 151–55, 159, 192; dispute with US Army Corps of Engineers, 166–67, 177–78; endorsement of wide-levee project, 180–81, 182, 183, 184; environmentalists targeting, 161–62; flood control projects of, 144, 147–48, 171–72, 180, 188, 197–98; General Assembly of Indiana and, 138, 185–86; Izaak Walton League and, 160, 167; "Memorandum of Understanding," 161; public's response to, 143–44; Roorda created, 138, 143; *The Times* criticizing, 149–50; West and, 156n10; Wide Levee Planning Committee of, 182

Kankakee River Basin Report, 147–48, 155

Kankakee River Basin Study, 146–47,

156n10

Kankakee River Basin Task Force, 148, 170, 191

Kankakee River Improvement Company, 94–96, 98

Kankakee Valley, 8, 18; landowners of, 83, 89–90, 91, 134–35; US Biological Survey of, 122

Kankakee Valley Draining Company, 53–54, 55–56, 58, 82, 139n6; Cass, G., and, 60–61, 71; ending of, 63–64; farmers protesting, 92–93; Hannah defending, 61–62; Whitten on failure of, 88, 129

Kankakee Valley Improvement Company, 91, 96, 98

*Kankakee Valley Post-News* (newspaper), 154–55

*The Kankakee Yesterday and Today* (newspaper), 170

Kenricks, John, 54

*Kentland Gazette* (newspaper), 35, 38–39, 42

Kimball, James, 73, 76

Kissimmee River, 133, 164, 200–201

Kite, Joseph, 19

KRBC. *See* Kankakee River Basin Commission

Kruse, bill, 190–91

labor, 4–5; ditches and, 31, 51; money and, 7, 22; value of, 6, 204, 209n26

Lake County, Indiana, 27, 159; ditches constructed along, 90; *Standard History of Lake County* (1915), 18; Sumava Resorts in, 135, 137, 179; Williams Levee in, 183

Lake County Soil and Conservation Service, 143

Lamborn, Jonathan, 56–57
land, 33; agricultural, 2, 124, 181; assessments for, 55–56, 57, 58, 59–60; Beaver Lake, 28, 36–37, 39–44, 47n33, 47n37, 51; ownership of, 5, 7; speculation, 41, 44; townships, 27, 28, 29–30, 36, 39, 42, 44, 103, 112. *See also* farmland
Landers, Franklin, 76
Landis, Charles, 90
landowners, 7–8, 102–3; Indiana, 1, 3; Kankakee Valley, 83, 89–90, 91, 134–35; Kankakee Valley Improvement Company, 91, 96, 98
land use: agricultural, 181; economics of, 205–6, 207; New Deal policies in, 15; permanent, 146; recreation, 145; The *Report on the Water and Related Land Resources of the Kankakee River Basin*, 144
Lant, Christopher, 22–23
LaPorte County, Indiana, 18, 95; farmland in, 97; meetings in, 58–60, 63; Mud Lake in, 92, 96, 97–98; wide-levee project in, 183
*The LaPorte Herald Argus* (newspaper), 59, 89
lateral ditches, 55, 65–66, 91–92, 95, 96, 100, 129, 186
Leopold, Aldo, 205–6
levees, 132, 145, 150, 161, 164; effectiveness of, 135; wide-levee project, 180–85, 187; Williams Levee, 183
Lieber, Richard, 116–17
Ling, F. E., 16
Locke, John, 4–9, 203
Luers, A. D., 143, 147, 151–55, 161; on KRBC, 163; *South Bend Tribune* interviewing, 159; US Army Corps of Engineers and, 161, 166

Lugar, Richard, 165, 169–70

maps, Meyers: chorography, 112–13; "Fundament of the Kankakee Marsh," 112; original marsh, 113, 114; vegetation, 113
Marble, Horace, 100–102
Marble–Powers Corporation, 138
Marble–Powers Ditch, 101–2
marsh hay, 15, 21, 113
Mast, Robert, 147
May, Allen, 35
McClellan townships, 36, 42, 44
McCool, Daniel, 164
McCorvie, Mary, 22–23
McEvoy, Arthur, 206
McFarland, Winfield, 146
McNamara, John, 151, 163
McWilliams dredging company, 101
Melton, Jody, 181–82, 183, 184, 186, 187, 197
"Memorandum of Understanding," 161
*Memories of Momence* (Morrison), 108n81
Menominee (chief), 5
Merchant, Carolyn, 8, 207
Meyer, Alfred H., 14–16, 21, 111–15, 122
*Michiana Farmers Journal* (newspaper), 143, 162
Milk, Lemuel, 36, 37, 40, 42–44
Miller Ditch, 97, 98
Mississippi River, 10n21, 22, 133; Great Flood on, 121; Izaak Walton League protecting, 120
Momence, Illinois, 2, 135; Board of Aldermen of, 81; *Memories of Momence*, 108n81; natural dam in, 60–61, 62, 65, 71. *See also* rock ledge
money, 35, 75, 95, 161, 163–64; for conservation, 116, 178, 187; for ditches, 87, 96–97; for drainage, 44, 53–54, 57, 66, 72, 88; labor and, 22; privately property

and, 5, 7, 38; value and, 3, 7
Morrison, Elizabeth, 108n81
Mud Lake, 92, 96, 97–98
Myers, Isaac, 39, 40

National Environmental Policy Act of 1969, 165, 189
national parks, 41, 44, 118, 191
National Wildlife Federation, 177
National Wildlife Fund, 178–79
natural dams, 14, 60–61, 62, 65, 71
*A Natural History of the Chicago Region* (Greenberg), 14
Natural Resources Conservation Service (NRCS), 206–7
natural world, 2, 3–4, 7–9, 117, 124, 203–4
navigability, of Kankakee River, 2, 165, 167
New, Jim, 150, 162
*New Castle Courier* (newspaper), 60
New Deal policies (1930), 15
*News Dispatch* (newspaper), 162, 190
Newton County, Indiana, 13; Historical Society of, 27; Milk owning, 42; *Standard History of Newton and Jasper Counties* (1916), 18–19
*Newton County Democrat* (newspaper), 58
*New York Times* (newspaper), 59, 71
Nicholas, Howard, 166
Nichols, Fay Folsom, 13
Niemeyer, Rick, 159–60, 207
NIRPC. *See* Indiana Regional Planning Commission
Nofsinger, W. R., 35
NRCS. *See* Natural Resources Conservation Service

*The Old Kankakee* (Izaak Walton League), 123
open ditching, 30–31, 51; Beaver Lake Ditch, 32–33, 34, 36, 42, 45
O'Reilly, James, 58
original marsh map, 113, 114
Osann, Ed, 177
outlet, for water, 31, 81, 99, 130, 137, 179
ownership, 3, 4, 5, 7, 28, 205; fraudulent, 31–32, 38, 41, 44; state, 39–40, 41. *See also* Bright, Michael G.

Packard, Marcus, 54
passenger pigeon, 123
Pelath, Scott, 197–98
permanent land use, 146
*Pioneer Hunters of the Kankakee* (Werich), 13
*Pittsburg Commercial* (newspaper), 59
Place, Dixon W., 91–94, 115, 202–3; at Porter County Court, 95; Tuesburg and, 103, 116, 117
Place Ditch, 92, 95–96, 98, 103, 106n38, 203; reconstruction of, 101; *Walkerton Independent* on, 107n38; Williams Ditch and, 100, 102
Pochedley, Elan, 207
political: consequences of drainage, 1; forces and the Kanakee River, 172–73
Porter, Albert G., 65, 76, 87
Porter County, Indiana, 18
Porter County Court, 62, 95
Potawatomi tribe, 5, 6, 7, 13, 20–21, 29, 207
Potter, Donald, 165
Power, Richard Lyle, 21
Powers, W. F., 101, 102
Pratt, Daniel, 41, 48n66
pre-emptors, of Beaver Lake, 37–38, 39, 40
preservation, 119, 165, 178, 207; Citizens for the Preservation of the Kankakee Basin, 184–85, 187; Leopold on, 206; of wetlands, 152, 153, 183

Prince, Hugh, 22, 23
private property, 4, 8, 23, 34, 205; money and, 5, 7, 38; public domain usurped by, 44, 51, 100
productive farmland, 3, 31, 32, 90, 115, 184, 204
*Program for the Future*, 178–80, 186
property rights, 51, 73, 74, 206; labor and, 4–5; private, 4, 7–8, 23, 34, 44; value of, 5–7, 72
public domain, 23, 29, 205; Bright stealing from, 36; private property usurping, 44, 51, 100
Puett, A. M., 32

Quayle, Dan, 169–70

railroads: Chicago and Eastern Illinois Railroad, 71, 74–75, 77, 80; development of Chicago, 90–91, 111; ecosystem of Kankakee Marsh changed by, 111
Reagan, Ronald, 207
recreation land use, 145
Reed, Earl H., 13–14
reflooding, 116
*The Regional News* (newspaper), 182
Reiger, John F., 118, 123
remonstrances, 64, 88, 92, 96, 97
removal, of rock ledge, 72–73, 77–79, 80, 81–83, 89, 129
*Rensselaer Republican* (newspaper), 170
*Report on the Improvement of the Kankakee River* (Campbell), 73, 139n6
The *Report on the Water and Related Land Resources of the Kankakee River Basin*, 144–45, 146
resolution, of US Senate and House of Representatives, 39, 48n66
restoration areas, for Kankakee Marsh, 125

Rice, James, 160
Ridenour, James, 183
Rinehart, B. F., 93–94
riparian claims, 33–35, 38–40, 42, 48n61, 74
Robertson, Bob, 163
rock ledge, in Momence, Illinois, 65, 66, 71, 76, 135; removal of, 72–73, 77–79, 80, 81–83, 89, 129. *See also* Board of Commissioners for the Removal of the Limestone Ledge in the Kankakee River
Rock Ledge Commission, 85n42, 129
Rolston, Holmes III, 203–4, 205, 208
Roorda, Walter, 138, 143, 144, 185–86
Roosevelt, Franklin, 114, 120, 164
Roosevelt, Theodore, 116, 117, 118, 204
runoff, from flooding, 179, 182
Rupel, C. F., 92–94, 202

Sackett, W. L., 104–5
sand deposits, 1, 98, 99, 148–49, 170, 171
Scherer, Nicholas, 5, 6
SCS. *See* US Soil Conservation Service
*Second Treatise of Government* (Locke), 6
sedimentation, 148, 152–53, 170–71, 199
SEG Engineers and Consultants, for Indianapolis, 183
SEG Master Plan, 184–86, 187
settlers, 5, 14, 18, 20–21, 29, 30, 55
Shelby, Indiana, 134, 135, 136, 149, 159
Shelby, William, 71–72, 76; Kankakee County Court and, 74, 75
sign, erected by Newton County Historical Society, 27, 28
Sirois, Emil, 75–76
Sisk, David, 78, 79
Slotkin, Richard, 203
snagging, of Kankakee River: clearing and,

152, 153, 154, 159, 160–62; Dovas on, 177–78; project for, 163–64, 165–66, 168–69, 171–73, 200
soil conservation practices, 152, 154
South Bend, Indiana, 13, 14, 183
*South Bend Register* (newspaper), 59, 63
*South Bend Tribune* (newspaper), 63, 149, 159, 160, 169
Spann, N. A., 154
speculators, 3, 22, 28, 32, 33, 38, 62
sportsmen, 15, 21, 104, 105, 119; *American Sportsmen and the Origin of Conservation* (Reiger), 118; commercial hunters and, 123
Staehle, William, 151
*Standard History of Lake County* (1915), 18
*Standard History of Newton and Jasper Counties* (1916), 18–19, 58; on Bright, M., 33–34
Starke County, Indiana, 17, 92, 96–97, 101, 149, 190; Yellow River, 98–100
*Starke County Democrat* (newspaper), 95, 101
*Starke County Republican* (newspaper), 96–97, 103, 129
State Budget Agency, Illinois, 161
State Budget Committee, for Indiana, 183
state legislature, of Indiana, 35, 118, 139
state line: Beaver Lake near, 3; confrontations at, 108n81; ditches near, 31; dredging near, 82–83, 102–4; farmers on, 171; Illinois, 3, 13, 72, 137; Indiana, 72, 73, 104; Kankakee River and, 2, 14, 101; land assessments near, 57; rock ledge removal around, 89
state ownership, 39–40, 41
steam dredge boats, 23, 66, 89–90, 96, 101, 104–5, 132; hunting impacted by, 118; opposition for, 105, 108n81; Williams Ditch and, 102, 103
Stevens, Wallace, 207–8
St. Joseph County, Indiana, 17, 54, 55, 57, 58, 60, 97–98
straightening, of Kankakee River, 1, 52, 53, 55, 91, 102–3; Campbell's proposal for, 65–66, 88–89, 92, 129; environmental change caused from, 111–12, 115; General Assembly of Indiana on, 104; Marble Ditch and, 101; Place Ditch and, 95–96, 98; Whitten on, 80, 99, 131, 134
Strausberg, Stephen, 32
Sumava Resorts, 135, 137, 179
surveyors: of Beaver Lake, 28–29, 45n8; pre-emptors and, 37
Swamp Land Acts of 1850, 22–23, 30, 54, 65, 88, 93

*Tales of an "Old Border Town" and along the Kankakee* (Burroughs), 19, 117
*Tales of a Vanishing River* (Reed), 13–14
*Tales of Kankakee Land* (Bartlett), 16–17
Tanke, William, 163
Test, Charles, 20, 28, 29, 39–40
Themer, Robert, 171–72
*The Times* (newspaper), 133–34, 149–50, 151, 165, 179
"Time to Call a Halt," 119
topographical features, of Kankakee Marsh, 14–15, 21
Townsend, Washington, 40–42, 44, 49n68
townships, 27, 28, 29–30, 39, 103; map of, 112; McClellan, 36, 42, 44
tree removal, 154, 168–69, 170, 171
Tuesburg, Charles, 96, 97, 100–101, 103, 116, 117
*Twentieth Century History of La Porte County* (book), 92

*Two Treatises of Government* (Locke), 4

unconstitutionality, 57–58
United Nations, 190–91, 203–4
Urban, Michael, 21
US Army Corps of Engineers, 85n42, 133–36, 144, 149, 155; environmental damage caused by, 164; Final Environmental Statement of, 172; Fithian and, 137; KRBC disputing with, 166–67, 177–78; Luers and, 161, 166; permit for clearing and snagging from, 163–64; permit for wide-levee project, 181–82; sued by Illinois, 165–66, 167; tree removal and, 168–69
US Biological Survey, 121, 122
US Board of Engineers for Rivers and Harbors, 99
US Congress: Beaver Lake and, 39, 42; Bright, M., and, 35–36; House of Representatives, 39, 40, 41, 56, 62, 63, 108n69; Swamp Land Acts of 1850, 22–23, 30, 54, 65, 88, 93; US Fish and Wildlife Service prohibited by, 191–92; US Land Office and, 28
US Department of Agriculture, 144, 206
US Fish and Wildlife Service, 122, 166, 187–88, 189–90; prohibited by US Congress, 191–92
US Land Office, 28
US Senate, 39, 63; Beaver Lake lands bill and, 41, 42, 47n33, 49n79; Drainage Act of 1869, 55–57, 88; Test and, 40
US Soil Conservation Service (SCS), 133, 137, 147, 177, 181, 198

Valparaiso, Indiana, 59, 66, 145, 148, 182, 184–85
value: of conservation, 3; economic, 2–3, 8, 119; Hofmann on, 124; of labor, 6, 204, 209n26; money and, 3, 7; of natural world, 3–4, 7–8, 124, 203–4; of property, 5–7, 72
vegetation, 131, 152, 172, 199; clearing of, 167, 168, 171; Meyers map of, 113; prairie, 45; swamp, 122
*Vidette-Messenger* (newspaper), 115, 116, 120–21, 146, 173

Walker, John C., 51–54, 81–82; Hannah and, 55
*Walkerton Independent* (newspaper), 92–93, 94, 107n38
Wallace, Byron, 147, 161
waterfowl, 124–25; on Beaver Lake, 14, 19–20, 44, 117; channelization reducing, 133; on Kankakee River, 114
Weir, Connie, 169
Werich, J. Lorenzo, 13
West, Roy M., 146–47, 156n10
Wetland Reserve Easement Program, 206–7
wetlands: Dickens on, 21–22; General Assembly of Indiana and, 30; inland, 1; preservation of, 152, 153, 183; sales of, 20; turned to farmlands, 23
White House, 116, 117
Whitten, William, 84n23, 89, 203; on Board of Commissioners for the Removal of the Limestone Ledge in the Kankakee River, 73, 75, 76, 79, 81–82, 85n42; on Campbell's general plan, 129–32, 140n24; on channelization, 200; *Chicago Tribune* quoting, 80; on failure of Kankakee Valley Draining Company, 88, 129; on flooding, 151; glacial drift deposits discovered by, 104; report on marsh drainage, 87–88; on

straightening of Kankakee River, 80, 99, 131, 134
Wide Levee Planning Committee, KRBC, 182
wide-levee project, 180–85, 187
wildlife: channelization reducing, 133; endangered, 8, 166; loss of, 120, 123; passenger pigeon, 123; refuge, 118; US Fish and Wildlife Service, 122, 166, 187–88, 189–90
Williams, L. R., 102
Williams Ditch, 102, 103–4
Williams Levee, 183
Willow Slough Fish and Wildlife Area, 125
Wisnieski, Pat, 202, 203
Wolf, Lee, 160
Woods, Bartlett, 54

Yellow River, 98, 100, 153, 162; Elsbree Ditch on, 99; flooding of, 136, 149, 159–60; Kankakee River Basin and Yellow River Basin Development Commission, 198–99; levee for, 183
Yellowstone National Park, 41, 44, 191

# ABOUT THE AUTHOR

MICHAEL DOBBERSTEIN IS A RETIRED ASSOCIATE PROFESSOR OF English, Purdue University Northwest. His teaching and research interests were in art and poetry. After retirement, he became interested in environmental history, with a special focus on the wetlands of northwest Indiana. He has published articles on the history of draining these wetlands, and continues to research and speak to local groups about them. The current work is the fruit of six years of research conducted in Indiana historical societies, state archives, and newspaper archives.

www.ingramcontent.com/pod-product-compliance
Lightning Source LLC
Chambersburg PA
CBHW051635230426
43669CB00013B/2312